VOLATILE CAPITAL FLOWS

TAMING THEIR IMPACT ON LATIN AMERICA

Edited by
Ricardo Hausmann and Liliana Rojas-Suárez

Published by the Inter-American Development Bank
Distributed by The Johns Hopkins University Press

Washington, D.C.
1996

The authors would like to thank Guillermo Calvo for many useful discussions and Antonio Spilimbergo for background work while these papers were being written; Deborah Davis and Michael Treadway for their contributions during the editorial process; and Erik Wachtenheim and Iván Sergio Guerra for their very able research assistance.

The views and opinions expressed in this publication are those of the authors and do not necessarily reflect the official position of the Inter-American Development Bank.

**Volatile Capital Flows
Taming Their Impact on Latin America**

© Copyright 1996 by the Inter-American Development Bank

Inter-American Development Bank
1300 New York Avenue, N.W.
Washington, D.C. 20577

Distributed by
The Johns Hopkins University Press
2715 North Charles Street
Baltimore, MD 21218-4319

Library of Congress Catalog Card Number: 95-82071
ISBN: 1-886938-04-0

INTRODUCTION

The Inter-American Development Bank's conference on international capital flows, on which this volume is based, took place in Jerusalem in April 1995. Had it been held only a few months earlier, the focus of the conference would surely have been quite different. The emphasis would have been on the excessive capital inflows that continued well into 1994, and presented many Latin American countries with some difficult policy trade-offs. Instead, the region in early 1995 unexpectedly found itself in a very different set of circumstances. This was yet another episode in the history of international capital flows, which has been characterized by waves of optimism and pessimism, boom and crisis—invariably amplified by swings in investors' perceptions that are less than fully justified by changes in the underlying fundamentals.

In September 1994, when the Bank decided to hold this conference, we already had some sense of an approaching turning point in the cycle of international capital flows. Rising interest rates in the industrial economies were the principal signal that a change was coming. We expected that the conference participants would focus on the longer-term issues of how to manage large capital inflows in a volatile macroeconomic environment, rather than on the exigencies of a short-term crisis. But a short-term crisis was precisely what erupted in the months leading up to the conference, and therefore captured much of the attention of the participants. The reader will find, however, that preoccupation with the immediate crisis did not prevent the panelists from drawing some lessons for Latin America and its future integration into world financial markets, particularly from the experience of abundant capital inflow that preceded the crisis.

One lesson is that whatever external shocks were in store for Latin America, economic liberalization, to which many countries of the region had already committed themselves, remained a valid strategy for establishing the long-term viability of the region's economies. Indeed, just a few months after the crisis, prices of Latin American equity and Brady bonds recovered sharply, signaling the renewed confidence of investors in the region's prospects.

We also learned what an enormous potential force world financial markets are for promoting development. The recovery of investment and growth in Latin America in recent years would have been much more difficult had it not been for the renewal of capital inflows in the early 1990s.

Yet we also witnessed a vivid demonstration of how volatile those markets can be, and of the consequences that follow when capital inflows fall or reverse themselves. We have seen how capital market volatility extends across countries—how a crisis in one country can quickly spread to others, despite different circumstances and different policies—but also how countries can to some extent insulate themselves from the shock.

The participants in the conference were asked to consider the following questions, among others:

- Can and should the countries of Latin America influence the magnitude and composition of the capital inflows they receive?
- Was the recent turmoil in global financial markets a rational response to some change in the region's underlying fundamentals, or did it reflect a degree of overreaction?
- What is the appropriate response of the governments of the region to volatile capital flows? How can they ensure that their economies adjust to the reduced availability of international capital without jeopardizing important domestic economic and social goals?
- What has been the impact of the crisis on domestic banking and financial systems? And what is the appropriate response of the international financial institutions? Has the ad hoc response observed so far proved adequate? If not, should international monetary cooperation be enhanced in such a way as to ensure a better and more timely response to future crises?
- Does the greater mobility of international capital have any implications for domestic policy? Does it strengthen the case for fixed or for floating exchange rates? Does it place new demands on fiscal policy?

The commentators on the papers delivered at the conference addressed these and other questions not only with great insight born of their extensive experience as senior economic and financial policymakers and as seasoned market participants, but also with a profound appreciation of the changes that have swept through Latin America in recent years. It is ever more widely recognized that the reforms undertaken by the countries of Latin America over the last decade exceed in scope anything seen in the region in the past half century. The economic fundamentals in countries that have carried through those reforms remain solid, and there is a manifest political will to deepen and extend the reforms.

On behalf of the Inter-American Development Bank, I thank the conference participants for their recognition of the progress made by Latin America in its effort to integrate itself into the world economy, and for sharing with us their expertise and advice.

Enrique V. Iglesias
President, Inter-American Development Bank

CONTENTS

PART I
The Macroeconomics of Capital Flows to Latin America: Experience and Policy Issues
Michael Gavin, Ricardo Hausmann, and Leonardo Leiderman ... 1

Commentary
Michael Bruno 41
Domingo Cavallo 45
Guillermo Perry Rubio 49
Lawrence H. Summers 53

Conclusion to Part I
Jacob Frenkel 58

PART II
Achieving Stability in Latin American Financial Markets in the Presence of Volatile Capital Flows
Liliana Rojas-Suárez and Steven R. Weisbrod 61

Commentary
L. Enrique Garcia 93
David Mulford 97
Arturo C. Porzecanski 102

Conclusion to Part II
Jacob Frenkel 105

Part I

THE MACROECONOMICS OF CAPITAL FLOWS TO LATIN AMERICA

EXPERIENCE AND POLICY ISSUES

Michael Gavin, Ricardo Hausmann, and Leonardo Leiderman

Introduction

After a hiatus of nearly a decade, the flow of international capital to Latin America resumed at the beginning of the 1990s. Having adjusted to scarcity, Latin America soon found itself faced with the relatively unfamiliar challenge of managing an abundance of international capital. Although often characterized as a problem, the renewed flow of capital was also widely and no doubt correctly perceived as an international vote of confidence in the liberalization and stabilization measures undertaken by most Latin American economies in preceding years, and as a valuable opportunity to deploy international savings to promote development of the domestic economy. The most serious concerns about the capital flows stemmed not from the inflows themselves, but from the possibility that they might, for reasons internal or external to the recipient countries, abruptly slow or even reverse themselves, thus forcing a potentially abrupt and painful macroeconomic and financial adjustment.

These concerns were proven legitimate during the course of 1994 when, after a series of domestic political shocks in Mexico and a substantial tightening of United States monetary policy, the rate of capital flow to Mexico slowed sharply and, toward the end of the year, reversed itself. By the beginning of 1995, the reduction in capital inflows became more generalized, contributing to financial instability in much of the region and to major financial crises in Mexico and Argentina.

This paper reviews recent experience with international capital flows in Latin America, and discusses the policy issues that surround them. The paper is predicated on three basic premises:

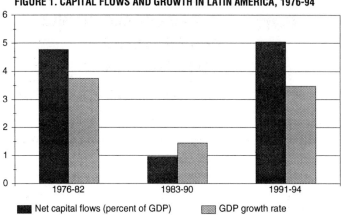

FIGURE 1. CAPITAL FLOWS AND GROWTH IN LATIN AMERICA, 1976-94

- capital flows to the region are an important source of macroeconomic disturbance;
- capital flows are very volatile, and the large fluctuations in these flows are due in substantial part to factors external to Latin America; and
- the fluctuations require a policy response. Policy should respond to sudden inflows or outflows of capital. Policy can also create institutions and regimes to reduce the economy's vulnerability to volatile international capital.

Capital Flows Are Important

As illustrated in Figure 1, there is a strong empirical correlation between economic growth in Latin America and flows of capital to the region. During the inflows episode of 1976-1982 (henceforth referred to as "the 1970s"), Latin America's economy grew nearly 4 percent a year, while receiving net capital flows of nearly 5 percent of GDP. During the period of capital scarcity that began in mid 1982 and ended around 1990, growth fell to less than 1.5 percent a year, while net capital flows to the region declined to less than 1 percent. And in 1991-1994, growth of about 3.5 percent a year was accompanied by capital inflows amounting to roughly 6 percent a year.

This relationship is of course partly noncausal, since both growth and capital inflows depend largely upon similar aspects of the policy and nonpolicy economic environment; and the causality that does exist clearly runs in more than one direction, since international investors tend to search for regions in which rapid growth can be expected. But the

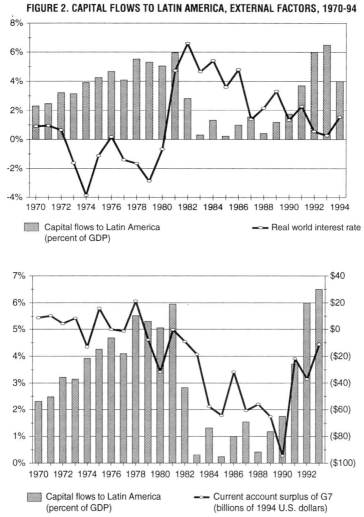

FIGURE 2. CAPITAL FLOWS TO LATIN AMERICA, EXTERNAL FACTORS, 1970-94

figure nevertheless illustrates what theory and Latin American experience would lead one to expect: capital inflows can provide a strongly expansionary impulse to the domestic economy, and sharp reductions in the rate of inflow can be strongly contractionary, at least in the short run.

Capital Flows Are Volatile and in Large Part Exogenous to the Region

Figure 2 shows the behavior of net capital flows to Latin America since 1970, and compares it with movements in world real interest rates. Two lessons emerge from this comparison. First, the flow of capital to Latin

America is very volatile. From a peak of about 6 percent of GDP in 1981, net capital flows to Latin America abruptly dropped to roughly nil in 1983, and stayed close to that level until around 1990, when they increased almost as abruptly to about 4 percent of Latin America's GDP in 1991 and to 6 percent of GDP in 1993 and 1994.[1]

Second, the flow of capital to Latin America is very largely determined by developments in the world economy that are essentially exogenous to Latin America. Figure 2 shows that net flows to the region are highly correlated with the world interest rate, over which economic developments in Latin America have little influence. The turning point of each major phase of the capital flows cycles experienced since the 1970s can be correlated with a substantial movement in world interest rates.[2]

A comparison of capital flows to Latin America and the current account surplus of the major industrialized economies also highlights the link between fluctuations in the availability of international capital and realized capital flows to the region. Figure 2 shows that capital flows to Latin America were low in the 1980s, when the large industrial countries were running substantial current account deficits, and that the surge of inflows to Latin America during the 1990s was associated with a sharp reduction in the G7 current account deficit. The most plausible interpretation of this figure is that shifts in the industrial countries' saving/investment balance are reflected in fluctuations in realized capital flows to Latin America (and, of course, elsewhere).

This does not mean that capital flows are unaffected by the domestic policy environment. In Latin America during the 1990s, countries that liberalized aggressively, reduced inflation, and maintained or created an open trading and financial system received larger capital flows than those that did not. It does mean that whatever the domestic policy environment, international capital flows are likely to remain an important mechanism through which shocks to the industrialized economies are transmitted to Latin America. Those shocks will require important macroeconomic adjustments, for which the region needs to prepare.

---

[1] Figure 2 visually understates the rapidity of the drop in international capital flows in this period. The flows remained large through August 1982 but fell very quickly to negligible levels after Mexico announced its inability to service its international debt according to schedule.

[2] Calvo, Leiderman, and Reinhart (1993) find that foreign factors accounted for 30 to 60 percent of the variance in real exchange rates and reserves. Chuhan, Claessens, and Mamingi (1993) and Fernandez-Arias (1993) also find that external factors explain a large proportion of capital flows to Latin America. In the discussion that follows, we focus primarily upon external shocks to capital flows.

Capital Flows Are a Policy Issue

A sudden shift in the availability of external finance will require important macroeconomic adjustments. A reduction in capital flows will typically generate an increase in domestic interest rates and a decline in asset values. This will help secure the reduction in domestic expenditure that is required for consistency with the new, lower supply of foreign finance, but it will also have adverse implications for domestic investment, and could generate a sharp contractionary impulse to the economy. The reduction in capital flows will also require a depreciation of the real exchange rate, with implications for employment and production in the tradables and the nontradables sectors, and creating the need for costly reallocations. Through its effect on the exchange rate and the balance of payments, the shock to capital flows will affect domestic prices and monetary aggregates, with potentially adverse implications for inflation.

Many of these adjustments are normal and desirable aspects of adjustment to the changed external environment. Policymakers cannot, however, afford to assume that adjustment will be trouble free. In world markets, adjustment to sharp fluctuations in capital flows is a policy issue because:

- International financial markets are incomplete, and do not provide insurance against possible risks associated with fluctuations in the magnitude of the flows.
- International financial markets may be subject to fads, bubbles, or contagion effects, in which international investors make sudden revisions about prospects for an economy that are unwarranted by underlying fundamentals. Because many of these fundamentals depend upon the degree of an economy's access to international capital markets, there is the possibility of multiple equilibria—sudden shifts in market sentiment can be self-fulfilling.
- Labor and product markets may be subject to externalities that distort the process of adjustment to changes in capital flows, making the private response suboptimal.[3]

---

[3] Business cycle fluctuations associated with fluctuations in capital flows may involve such externalities: see Blanchard (1987) and Cooper and John (1988). The sectoral reallocations induced by the real exchange rate fluctuations that often accompany changes in the rate of capital flow may also be subject to such externalities. Mussa (1982) describes a number of imperfections in the adjustment process. In Gavin (1993), there is too much unemployment associated with sectoral adjustment of the economy, providing a rationale for policy to "lean against the wind" in response to shocks.

- Domestic financial markets may be subject to information or policy-generated imperfections that cause them to intermediate capital inflows suboptimally, thus increasing the economy's vulnerability to subsequent reductions in the rate of inflow.
- Sharp fluctuations in international capital flows may interfere with the effectiveness of other government policies, including attainment of price stability and management of aggregate demand.

These considerations raise two related policy problems. The first is the formulation of appropriate policy responses to increases and decreases in international capital flows as they materialize: for example, should fiscal policy expand or contract when international capital becomes more scarce? The second is the design of appropriate institutions and policy regimes to reduce the economy's vulnerability to fluctuations in these flows, and to minimize the adverse effects of their volatility on the real economy. Here we might ask: what exchange rate regime provides a better adjustment to sharp changes in the availability of foreign capital? Can the regulatory environment within which domestic and international investors operate affect the magnitude of such shocks, and their impact on the economy?

The remainder of the paper takes up these and related issues in more detail. Section II lays out the points of macroeconomic contact between international capital flows and the macroeconomy, with the aim of highlighting aspects of the transmission mechanism that create policy problems, and ways in which policy can alleviate the problems. We also examine the recent experience of Latin America with capital inflows, focusing on the most recent inflows episode and the lessons that emerge from the response to the reduction of those flows toward the end of 1994. Section III brings lessons from these discussions to bear on a discussion of specific policy questions surrounding the flows.

Capital Flows and the Macroeconomy

In this section, we outline the ways in which capital flows and policy interact to determine macroeconomic outcomes, in order to highlight policy issues that are discussed in the subsequent section. The discussion is organized around a simple schematic approach that is summarized in the chart on the opposite page.

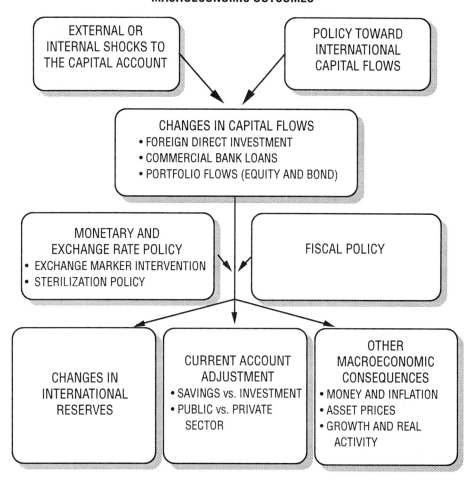

We begin with a discussion of sources of shocks to the availability of international capital. We then discuss how such shocks and policy toward capital inflows interact to determine the magnitude and composition of the flows. Next we discuss how the realized capital inflows interact with monetary, exchange rate, and fiscal policy to determine macroeconomic outcomes; for example, whether the capital inflows finance reserve accumulation or current account deficits, or whether the current account deficits that do emerge are generated by increased investment or lower saving. Finally, we consider other macroeconomic consequences of the flows, including effects on money and prices, on interest rates and asset prices, and on growth and real activity.

The diagram does not, of course, fully capture all the relevant macroeconomic interrelations, and variables in the upper part of the diagram are not, in general, exogenous to those listed in the lower part. For example, exchange rate intervention and sterilization policy may well affect the magnitude of the change in capital flows to which the economy is subject. But the framework helps to clarify certain important channels of influence, and provides a guide for the discussion that follows.

Shocks

Observed fluctuations in the rate of capital inflow or outflow may be due to either internal or external factors. Domestically generated shocks may come from dislocations to the real economy, such as a transitory productivity shock or a sudden shift in macroeconomic policy that generates an abrupt change in the balance between domestic savings and investment. Domestically generated shocks may also come from financial markets, as when news about political outcomes alters domestic and foreign investors' confidence in the economy and induces capital inflows or capital flight.

External factors can affect an economy directly, as when interest rates in the world economy change, thereby increasing or reducing the relative attractiveness of domestic assets. They may also be the result of contagion, or bandwagon effects, as when adverse news about one country's creditworthiness alters international investors' perceptions about the creditworthiness of others.[4]

In the 1990s, external shocks have been an important determinant of capital flows to Latin America and other emerging market economies. Arguably the most important of these factors has been changing cyclical conditions in the major industrial countries, which largely determine world interest rates and economic activity. The correlation of capital flows to Latin America with world interest rates and with the savings/investment balance of the major industrial countries has already been noted. Low world interest rates affected the attractiveness of investment in Latin America both directly and by raising the creditworthiness of those many emerging market economies that are net debtors.

Fluctuations in international capital flows may also arise from changes in the portfolio preferences of international investors, which may in turn be caused by changes in the regulatory environment in which

---

[4] Calvo and Reinhart (1994) provide empirical evidence on such contagion effects in Latin America.

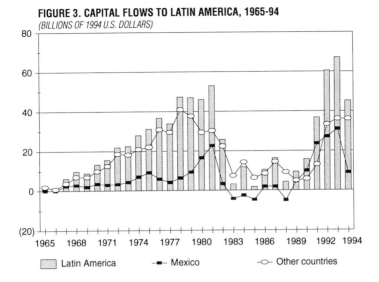

FIGURE 3. CAPITAL FLOWS TO LATIN AMERICA, 1965-94
(BILLIONS OF 1994 U.S. DOLLARS)

these investors operate, or in information technologies. Shocks to foreign investment may also result from actual or anticipated changes in regional trade arrangements; for example, the North American Free Trade Agreement may have created a substantial incentive to invest in Mexico to produce for the North American market.

Whatever their cause, shocks to the availability of international capital can be very big, relative to the economies of Latin America. Figure 3, showing recent fluctuations in capital flows to the region, reveals important differences between earlier and later capital inflows. In 1993 and 1994, capital inflows were even larger than in the late 1970s and early 1980s. The recent surge in the flow of capital was also substantially more abrupt than in the previous episode, and may be of shorter duration.

Indeed, in the years leading up to the 1982 onset of the debt crisis, net capital flows to countries other than Mexico had been declining. For many countries, the debt crisis of the 1980s was generated as much by domestic capital flight as by inappropriately high levels of net borrowing—a fact that distinguishes the most recent episode from the earlier one.

Figure 3 also illustrates that the inflows episode of the 1990s is to a very large extent a story about Mexico, and to a lesser extent Argentina (not shown in the figure). Mexico's experience with international capital markets in the 1982-1989 period of capital scarcity was harsher than typical; in many of those years, Mexico experienced net capital outflows, unlike the rest of the region.

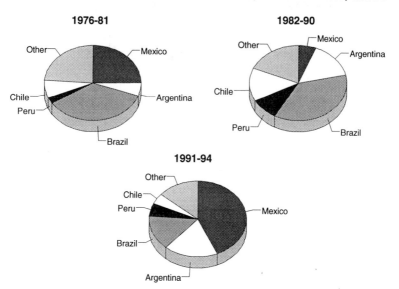

FIGURE 4. CAPITAL INFLOWS TO LATIN AMERICA, MAJOR RECIPIENTS, 1976-94

But after that experience, renewed inflows began earlier in Mexico than in most of the region, presumably because Mexico resolved its debt problems, stabilized inflation, privatized aggressively, and liberalized its economy relatively early. The recovery of international investment in Mexico was dramatic, and in the early years of the inflows episode, flows to Mexico accounted for well over half of the capital flows to the region, greatly in excess of Mexico's share in the regional economy.

For the 1991-1994 period as a whole, as shown in Figure 4, Mexico accounted for nearly 45 percent of the inflows, despite the sharp decline in flows to Mexico during 1994. Argentina accounted for another 18 percent of the inflows, somewhat larger than her size in the regional economy. Taken together, Mexico and Argentina accounted for over 60 percent of the net flows to the region, while Chile and Brazil—heavy borrowers during the 1970s—figured much less prominently.

Measured relative to the size of the recipient country's economy, a somewhat different picture emerges, as shown in Figure 5. Relative to GDP, Peru was the largest recipient of capital flows, with inflows amounting to roughly 13 percent of GDP, during the 1991-1994 period. Mexico still ranked very high, with inflows of nearly 8 percent of GDP, but inflows to Argentina averaged only about 4.5 percent of GDP, not much more than the average for the continent.[5]

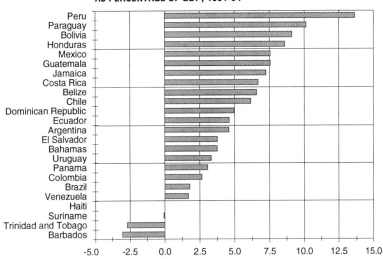

FIGURE 5. CAPITAL FLOWS TO LATIN AMERICAN COUNTRIES, AS PERCENTAGE OF GDP, 1991-94

Of the larger countries, Brazil, Colombia, and Venezuela have experienced much smaller capital inflows, in comparison with their size, than other major countries in the region. In Brazil and Venezuela, it seems that macroeconomic turbulence—stabilized in Brazil during 1994 and still ongoing in Venezuela—made international investors wary. Colombia, on the other hand, has made substantial efforts to insulate its economy from volatile international capital flows, and the relatively low rate of flow into the country during 1991-1994 may be seen as evidence that the authorities were at least partially successful in doing so.

This comparison illustrates the important point that policy and external shocks are not independent, but rather interact in determining capital inflows. In Latin America during the 1990s, nearly all countries participated in the capital inflows associated with low world interest rates, but countries that were on good terms with their international creditors, that had stabilized inflation and brought fiscal deficits under control, and that maintained open trade and financial regimes, tended to receive more international capital than countries that had not. The rela-

---

[5] Some care needs to be taken in interpreting these comparisons, since the required computation of U.S. dollar GDP is highly sensitive to the exchange rate; and when the exchange rate is overvalued, the ratios will tend to understate the magnitude of the flows compared with the long-run sustainable levels of (dollar) GDP.

tively low rate of capital flow to Chile and Colombia despite generally good macroeconomic fundamentals is limited evidence that attempts to moderate the rate of capital inflow may have some success, at least in the short run.

Composition of Capital Flows

The financial instruments used to effect the transfer of capital will have a substantial impact on the sharing of risk among residents of the recipient country and international investors, and may also have a significant impact on the volatility of international capital flows to which the economy is subject. Volatility is likely to be much higher if flows are of a short-term, purely speculative nature than it might be if flows primarily reflect foreign direct investments, which are presumably guided by medium- or long-term fundamentals. This is particularly true of short-term debt instruments, the principal of which becomes payable frequently. Equity investment is liquid, and can be sold by international investors upon short notice, but owners of equity bear much of the price risk associated with a generalized flight of capital from the economy. The composition of the flows also determines their degree of sensitivity to interest rate differentials.

The mechanisms through which capital was channeled to Latin America differed dramatically in recent years from those which typified the late 1970s and early 1980s. As illustrated in Figure 6, commercial bank lending was the dominant form of international intermediation during the 1977-1981 period, accounting for roughly two-thirds of total flows to Latin America.

In sharp contrast, commercial bank lending fell to barely more than 10 percent of the total in the 1991-1993 period, while foreign direct and equity investment rose to nearly 40 percent, and international bond issues increased to over 20 percent of the total.

These changes have important implications for the management of capital account shocks. The much higher share of equity investment means that foreign investors are exposed more directly to and share more completely in country risk than would be the case if commercial bank lending were still the dominant form of intermediation.[6] It has been estimated that roughly 20 percent of the Mexican stock market's capitalization was foreign owned at the beginning of 1994. Thus, of the more than

---

[6] This point would hold as well for long-term debt where there is significant price risk.

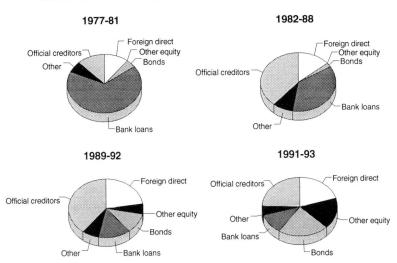

FIGURE 6. CAPITAL FLOWS TO LATIN AMERICA, BY TYPE OF INVESTMENT, 1977-93

Note: For 1977-81 and 1982-88, the value of "Other equity" is represented by the width of the line.
Sources: World Bank and IMF.

$140 billion reduction in the dollar value of the stock market that occurred between January 1994 and early March 1995, foreigner investors must have absorbed something approaching $30 billion.

At the same time, management of international liquidity or solvency problems may have become more difficult than it was in the 1980s, since there is at present no obvious counterpart to the bankers' committees that were formed in that period to negotiate with countries over the terms of debt rescheduling.

Policy can affect the composition of capital inflows in several ways. First, sustainable and credible economic policies are likely to result in less volatile capital flows, since international investors will be more willing to make relatively irreversible commitments to the economy. Second, tax and regulatory policies can be designed to reduce the attractiveness of short-term, speculative inflows, relative to those investments that imply a larger degree of commitment to the recipient economy.[7] In several countries, for example, there are taxes and restrictions on short-term foreign borrowing. And finally, various microeconomic policy measures can affect the attractiveness of the climate for foreign direct investment.

---

[7] Of course, countries should expect international investors to demand compensation, in the form of higher required returns, for the risks entailed by such commitments.

Balance of Payments Adjustment

Net capital inflows must by definition be matched by central bank reserve accumulation and/or increased current account deficits, while outflows must necessarily be accompanied by a loss of reserves and/or movement of the current account toward surplus. Domestic macroeconomic policy plays an important role in determining the pattern of balance of payments adjustment to sudden shocks to the capital account. The exchange rate regime is, of course, fundamental here.

Under fixed exchange rates, the exchange market intervention required to defend the parity in the aftermath of a positive shock to the capital account will lead to reserve accumulation in the early stages of an inflows episode, while current account deficits are likely to grow as the episode progresses. In the polar opposite case of a pure float, in which there is by definition no central bank intervention, any shock to the capital account must have as its counterpart an equal and opposite change in the current account balance. In the intermediate and more realistic case of flexible exchange rates with intervention, the amount of reserve accumulation is a policy choice. The more aggressive the reserve accumulation, the more thoroughly the authorities will insulate the nominal exchange rate from pressures generated by the capital flows.

Intervention will have important effects on the short-run adjustment to capital account shocks, as can be seen most clearly by contrasting the short-run adjustment under fixed exchange rates with that which would occur under a pure float. In the former regime, the capital inflows will lead to immediate reserve accumulation, which, with incomplete sterilization, will generate an increase in the domestic monetary base. This increase in liquidity will drive down interest rates and generate an expansion of bank credit—generating, over time and in the context of an output boom, the increase in domestic consumption and investment that is required to make the current account adjust to the new and higher availability of external finance. The opposite would, of course, occur after a sharp decline in the rate of capital inflow.

Under flexible exchange rates with no intervention, a shock to the capital account generates, by definition, no change in central bank reserves and creates, instead, exchange rate appreciation. This appreciation creates, through various mechanisms, a current account deficit in the context of depressed demand for domestic production.

In intermediate cases where exchange market intervention is neither dictated by the requirement to defend the exchange rate, nor ruled out by definition, the short-run response of the macroeconomy will depend upon the authorities' intervention behavior. That behavior will, in turn, depend in part upon how their reserve position compares with that which they consider optimal. If reserves are low at the beginning of an inflows episode, central banks may be expected to accumulate reserves in the initial stages, even if not required to do so by a formal exchange rate commitment. Over time, the tendency to accumulate reserves may decline.

The optimal degree of reserve accumulation also depends upon prospects for a reversal of the shock to the capital account. If there is a high probability that the capital that is currently flowing into the economy will want to leave in the near future, then it would be prudent to seek a higher level of reserves. Thus, when capital inflows are the result of very short-term and speculative investments, aggressive and prolonged intervention may be called for.

The capital inflows episode of the 1990s was accompanied by very substantial reserve accumulation, significantly larger than was observed during the 1970s. This was particularly true in the first years of the episode. In 1990, the accumulation of international reserves nearly matched the (still relatively small) net capital inflows, with the implication that the latter had no effect on the current account.

As the episode progressed, evidence of "reserve accumulation fatigue" materialized, as shown in Figure 7. For the region as a whole, the share of reserve accumulation in total capital inflows fell to 50 percent in 1991, 40 percent in 1992, and about 30 percent in 1993, with the implication that the capital inflows were associated with growing current account deficits. In 1994, the region's reserves actually declined, despite continued inflows.

This sharp decline in the rate of reserve accumulation is largely attributable to developments in Mexico and Venezuela, in which reserves in these countries were run down in 1994 to cope with macroeconomic and financial crises associated with large capital outflows. In other countries, reserve accumulation averaged about 75 percent of the capital inflows during the 1990-1992 period, declining to about 40 percent in 1993 and 1994.

Brazil accumulated reserves most aggressively during the recent inflows episode, as shown in Figure 8; indeed, its reserve accumulation reached almost 120 percent of total capital inflows during the period. Reserve accumulation also accounted for a large proportion of capital inflows in Chile, Colombia, Ecuador, and Peru. As noted above, reserve accumulation was quite low for Mexico and Venezuela, although this is largely attributable to events that transpired in 1994. In particular, in

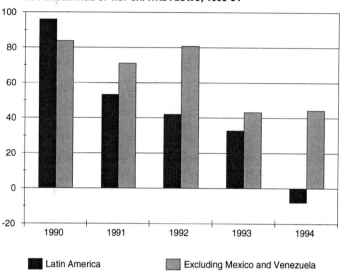

FIGURE 7. RESERVE ACCUMULATION IN LATIN AMERICA, AS PERCENTAGE OF NET CAPITAL FLOWS, 1990-94

Mexico the central bank accumulated reserves on a large scale during the first years of the inflows episode, but lost those reserves and more during 1994, when it responded to a sharp reduction in the rate of capital inflow by intervening in the foreign exchange market to maintain the peso within the limits of the band, while at the same time sterilizing the consequent reserve losses.

During a period of capital inflows, as reserves are accumulated, the central bank will be faced with the question of whether to sterilize the change in reserves, or instead allow it to affect the domestic money supply. If the money supply is allowed to adjust, (risk-adjusted) domestic interest rates will be driven toward world rates, the economy will adjust to the flows as described above, and the incentives for further capital flows will be reduced. Under sterilization, the macroeconomic adjustment to the capital inflows may be postponed, but because domestic interest rates will not be driven to world rates, the magnitude of the capital inflows may increase. The resultant quasi-fiscal losses, as the central bank issues high-interest domestic paper in exchange for low-interest reserve assets, can be substantial. Partly for this reason, attempts to sterilize the impact of large capital inflows have often been abandoned, or complemented in short order by attempts to directly reduce the rate of short-term capital inflow.

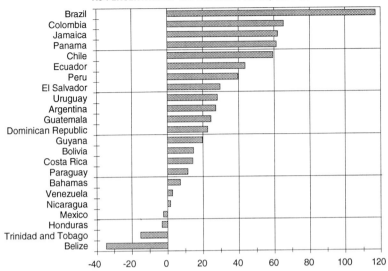

FIGURE 8. RESERVE ACCUMULATION IN LATIN AMERICAN COUNTRIES, AS PERCENTAGE OF NET CAPITAL FLOWS, 1991-94

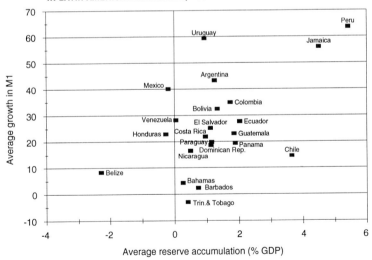

FIGURE 9. INTERNATIONAL RESERVES AND MONETARY GROWTH IN LATIN AMERICAN COUNTRIES, 1991-94

Note: The high-inflation countries Brazil and Suriname are excluded from this graph.

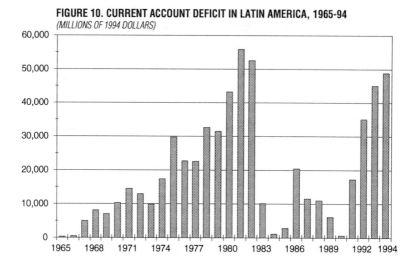

FIGURE 10. CURRENT ACCOUNT DEFICIT IN LATIN AMERICA, 1965-94
(MILLIONS OF 1994 DOLLARS)

There is some evidence that during the 1990s, countries found it difficult or undesirable to fully sterilize the reserve accumulation that was undertaken. As shown in Figure 9, countries that accumulated significant reserves tended also to experience rapid monetary growth. The relationship is, however, not particularly tight, suggesting that there was some scope for sterilization over the time period in question.

Chile stands out for having accumulated reserves in the amount of nearly 4 percent of GDP per year, while managing to maintain relatively moderate monetary growth.

Mechanisms of Current Account Adjustment

Under both fixed and flexible exchange rates, the current account will generally adjust at some point to a sustained capital inflow. Two key mechanisms are responsible for creating the movement toward current account deficit in response to capital inflows—a response that has been evident in Latin American economies during the 1990s.[8]

- The inflows will generally create a reduction in domestic interest rates and an increase in asset prices, thus promoting an increase in expenditure relative to production. Under fixed

---

[8] The following discussion is couched in terms of capital inflows for the sake of brevity. The case of outflows is largely symmetric, with potential asymmetries discussed in the text.

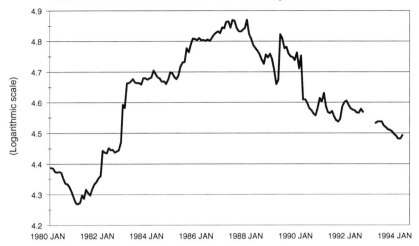

FIGURE 11. REAL EXCHANGE RATE IN LATIN AMERICA, 1980-94

Notes: Index uses GDP weights. An increase in the index represents a depreciation. The gaps represent missing data.

exchange rates, this is largely due to an expansion of liquidity and an associated increase in bank lending.
- Inflows will also create pressures for real exchange rate appreciation. Under flexible exchange rates, this is caused by nominal exchange rate appreciation. Under fixed exchange rates the real appreciation is created by domestic inflation. This may take longer to materialize than under flexible exchange rates, but in the long run the real exchange rate will respond in much the same way.

Current Account and the Real Exchange Rate

As indicated in Figure 10, the current account in Latin America moved dramatically toward deficit as capital inflows surged in the 1990s. The rate of reserve accumulation during 1991-1994 was substantially larger than in the 1970s. As a result, although the capital inflows were larger in the more recent period, current account deficits in the region as a whole have been somewhat smaller than in the previous inflows episode. They have nevertheless been large in aggregate.

Here, however, the regional average obscures important differences among countries. The graphs in Figure 12 show fluctuations in capital flows and current accounts for the eight largest countries in Latin

FIGURE 12. FLUCTUATIONS IN CAPITAL FLOWS AND CURRENT ACCOUNTS FOR THE EIGHT LARGEST LATIN AMERICAN COUNTRIES

(MILLIONS OF 1994 DOLLARS)

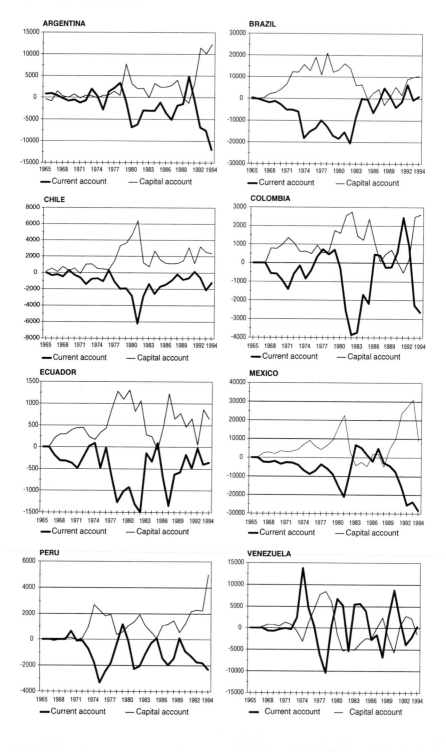

America. Most of these countries have experienced somewhat smaller current account deficits than in the previous inflows episode.

It is notable that Brazil, Chile, and Ecuador have experienced surpluses or small deficits in recent years, in very sharp contrast to the large deficits registered in the 1977-1981 period; and that while Colombia and Peru have recently experienced substantial current account deficits, they remain significantly lower than in the previous episode.

In Argentina and Mexico, however, external deficits are very large, in absolute amount and compared with those that the countries experienced in the years leading up to the debt crisis. For Mexico, the 1994 current account amounted to about 8 percent of GDP—substantially larger than in the years leading up to the debt crisis.

The large current account deficits were, in most countries, accompanied by substantial appreciation of the real exchange rate, as shown in Figure 11. For the region as a whole, the real exchange rate appreciated substantially in comparison with its very depreciated levels of the mid 1980s, but remained considerably more depreciated than in 1981 and 1982. For the region as a whole, there is a clear empirical association between the availability of international capital and the real exchange rate. The real exchange rate was very appreciated during the early 1980s, when international capital was plentiful, and depreciated dramatically when the flows of international capital slowed in the 1980s. It appreciated again during the years of international capital abundance in the early 1990s.

Again, the behavior of this regional average obscures important differences among countries. As shown in figures 13 and 14, the real exchange rate appreciated considerably during the inflows episode in Argentina, Colombia, Ecuador, Mexico, and Uruguay.[9] Chile experienced a smaller exchange rate appreciation, and in Brazil the exchange rate depreciated until the mid 1994 stabilization plan, after which it began to appreciate strongly. Note that substantial real exchange rate appreciation was experienced by countries operating under fixed as well as flexible exchange rate regimes.

It might be quite misleading to attribute the exchange rate appreciation solely or even primarily to the capital flows; in many countries, the timing of the exchange rate appreciation and associated changes in the current account match the timing of an inflation stabilization more

---

[9] The 1990 starting point is fairly arbitrary, and is chosen because it represents the beginning of the inflows episode for most countries. Because the Argentine exchange rate was in 1990 still influenced by the hyperinflation, Figure 13 probably overestimates the amount of any real overvaluation of the Argentine peso.

FIGURE 13. CHANGE IN THE REAL EXCHANGE RATE IN LATIN AMERICA SINCE 1990

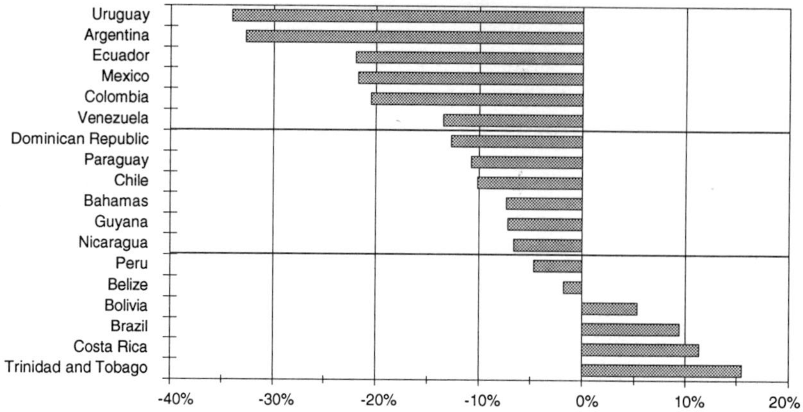

Uruguay	~-33%
Argentina	~-31%
Ecuador	~-23%
Mexico	~-22%
Colombia	~-21%
Venezuela	~-15%
Dominican Republic	~-13%
Paraguay	~-12%
Chile	~-10%
Bahamas	~-8%
Guyana	~-7%
Nicaragua	~-6%
Peru	~-4%
Belize	~-3%
Bolivia	~4%
Brazil	~9%
Costa Rica	~11%
Trinidad and Tobago	~16%

Note: Positive number signifies a depreciation.

closely than the timing of capital inflows. But whatever the causality, there is a clear correlation between changes in the real exchange rate and changes in the current account balance, as illustrated in Figure 14.

While this association between the current account and the real exchange rate does not establish causality, it does suggest strongly that the dominant shock to the current account was not a supply shock to the current account.

Most countries in the region experienced deteriorations in the current account during the 1990-1994 period. But the deteriorations were much larger in countries where the real exchange rate appreciated substantially than in countries where it did not.

Asset Prices

Along with appreciation of the real exchange rate, recipients of capital inflows in the 1990s experienced reductions in interest rates and booming asset prices. Figure 15 documents the collective behavior of the stock markets of four large Latin American economies: Argentina, Brazil, Chile, and Mexico.[10]

In real terms, this measure of equity prices stagnated during the 1980s, when international capital was scarce, and exploded during the inflows episode of the 1990s, rising by a factor of six over the course of

---

[10] The index is a geometric average of real, dollar equity prices, weighted by market capitalization in 1994. The U.S. producer price index is used to deflate the indexes.

FIGURE 14. THE REAL EXCHANGE RATE AND THE CURRENT ACCOUNT, 1990-94

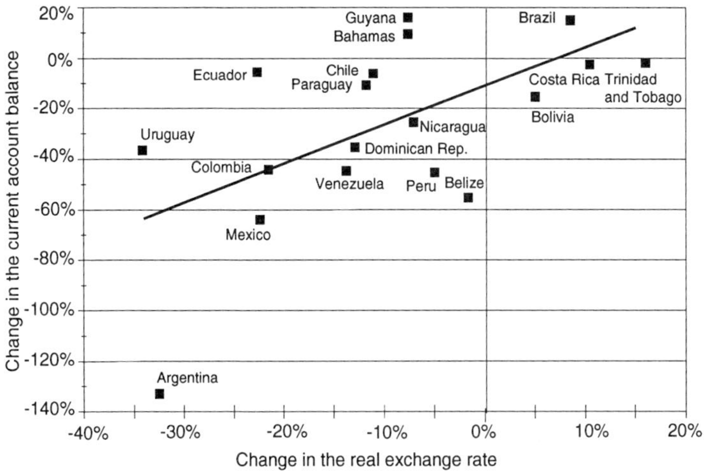

Notes: The change in the current account is scaled by the initial (1990) level of exports.
The line is the regression line indicated by the data in the scatterplot.

FIGURE 15. PRICES IN FOUR LATIN AMERICAN EQUITY MARKETS, 1986-95

only four years. All major countries except Venezuela[11] experienced very large increases in equity prices during the inflows episode.

The chart also shows that what capital markets give, they can take away; with the abrupt decline in the rate of capital inflow at the end of 1994, equity prices plummeted in Latin America. Declines have been widespread; as shown in Table 1, most countries experienced substantial

---

[11] At the end of 1994, Venezuela was well into a major economic and financial crisis associated with adverse shocks to oil income and the failure of a substantial portion of its banking system.

declines in domestic equity prices after the Mexican financial crisis. Here, however, it is noteworthy that Chile and Colombia have been affected to a much smaller degree than Argentina, Brazil, or Mexico itself.

This suggests that some aspect of these countries' former policy regimes provided insulation from the financial shock. Precisely which policy was responsible is difficult to say, because Chile and Colombia differ from Argentina in several ways:

- Their exchange rate policies are less rigidly tied to the achievement of specific nominal outcomes, and both countries have recently demonstrated a willingness to allow the nominal exchange rate to adjust in response to capital inflows.
- Both countries have made serious attempts to reduce the inflow of short-term capital during the inflows episode, and both countries sterilized roughly 60 percent of the inflows that did emerge. (Brazil sterilized aggressively during the 1991-1994 period as well, but the Brazilian policy regime changed dramatically with the mid 1994 stabilization.)
- Both countries recently experienced a significant and favorable shock to the terms of trade.

These financial market effects may create an important asymmetry between capital inflows and outflows. Large inflows are associated with increasing asset prices and ample domestic liquidity, which may lead to inappropriate lending decisions but will not create a crisis. Outflows, on the other hand, are often associated with declining asset prices, and under fixed exchange rates, may require a sharp contraction of the domestic money supply and, therefore, of commercial bank lending. Both of these can contribute to instability in the domestic banking system. In this environment, associated fears—whether rational or exaggerated—about the safety of the banking system can be highly destabilizing, and if they create a full-blown banking crisis, may undermine the government's fiscal and monetary policies.[12]

---

[12] Losses associated with declines in the price of internationally traded Argentine bonds were largely responsible for the 1995 collapse of a merchant bank in Argentina. Although the bank itself was relatively small, the collapse created confidence problems for a number of Argentine banks. More recently, the sharp contraction in the domestic financial system that has been imposed by the Argentine convertibility plan with capital outflows has generated interest rates so high as to call into question the ability of both the public and private sectors to service their debt.

TABLE 1. PERCENTAGE CHANGE IN EQUITY PRICES
(IN U.S. DOLLARS)

	1990- Dec. 1994	Dec. 1994- Mar. 3, 1995
Argentina	320	-35
Brazil	489	-32
Chile	345	-15
Colombia	575	-1
Mexico	481	-49
Venezuela	5	-5

Public vs. Private Adjustment

It is of major importance that the current account deficits that emerged in Latin America in the 1990s occurred despite a substantial improvement in fiscal performance in the region. As illustrated in Figure 16, central governments in the region registered an average fiscal deficit of more than 10 percent of GDP in 1988 and 1989. This was cut in half during 1990, and in 1991-1993 the deficit averaged less than 1 percent of GDP.[13]

There was also significant fiscal convergence during this period; the countries that made the most dramatic improvements in fiscal performance in the 1990s were those that began with the largest fiscal imbalances. This means that very few countries in Latin America had substantial deficits in the 1990s, and that the current account imbalances that emerged were predominantly the result of a growing gap between private savings and investment. We shall return to this point below.

Saving and Investment

A key question for policymakers is how the current account response to a change in capital flows is effected through changes in saving or domestic investment. To the extent that domestic savings and investment fall short, for various reasons, of the socially optimal level, policymakers may be concerned to ensure that savings does not fall too dramatically during periods of capital inflow, and that investment does not fall excessively when international capital becomes scarce.

---

[13] These fiscal data are not adjusted for inflation, which was declining rapidly during this period, or for the business cycle. Adjustment for these factors would reduce the magnitude of the fiscal adjustment, but it would nonetheless remain substantial.

In practice, capital inflows to Latin America have been associated with both increased investment and decreased savings, and conversely, the reduction in inflows during the 1983-1990 period was associated mainly with a decline in private and public investment. After the very sharp decline in capital flows to Latin America that began in 1982, the current account adjustment was achieved in very large part through a decline in both public and private investment, and investment remained extremely low by historical standards until well into the most recent inflows episode.

As the capital inflows of the 1990s began to make themselves felt in Latin American economies, domestic investment did recover from its extremely depressed levels of the late 1980s. As Figure 17 illustrates, in the region as a whole, (real) investment rose from less than 20 percent of GDP in 1989-1990 to over 22 percent of GDP in 1994. Setting Brazil aside, investment rose more dramatically, from roughly 18.5 percent in 1989-1990 to nearly 25 percent of GDP in 1994.

While an increase in domestic investment was thus an important cause of the current account deficits that materialized during the most recent capital inflows episode, savings also declined in most countries. The declines in national savings began at different times in different countries, and have been very pronounced in several countries.

The causes of these often very sharp declines in private savings are not fully understood. The decline in saving may have been due to wealth effects of the booming equity, land, and housing markets associated with the capital inflows. Or it may have been due to the relaxation of financing constraints attributable to the expansion of domestic bank credit associated with the monetary consequences of the inflows. Or domestic consumers may have been too optimistic about prospects for the future. Whatever caused these declines, their consequence was a substantial movement of the current account into deficit, thus utilizing the increased supply of international financing.

When current account deficits are associated with large fiscal deficits, there is little dispute about the desirability of reducing such deficits by bringing the government's budget toward balance. However, when the large current account deficits are associated with an imbalance of private saving and investment, as in the 1990s, the effectiveness and desirability of using fiscal or other policies to raise national saving are a matter of some dispute.

First, whether current account deficits generated by movements in private saving and investment pose a policy problem is itself a matter of substantial dispute. On the one hand, it can be argued that individu-

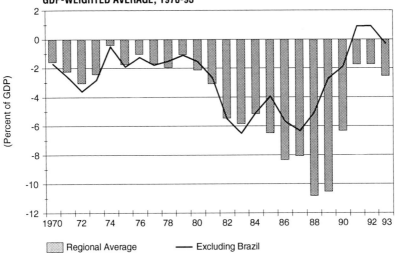

FIGURE 16. FISCAL DEFICIT IN LATIN AMERICAN CENTRAL GOVERNMENTS, GDP-WEIGHTED AVERAGE, 1970-93

Regional Average — Excluding Brazil

Note: Data are unavailable for Brazil for 1993 and 1994. It is assumed that the Brazilian deficit in those years equals the 1992 deficit. This probably understates the improvement in the regional fiscal balance in 1993 and 1994.

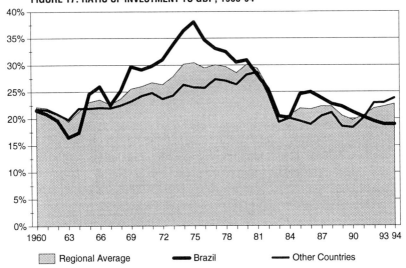

FIGURE 17. RATIO OF INVESTMENT TO GDP, 1960-94

Regional Average — Brazil — Other Countries

als know better than governments how best to arrange their intertemporal affairs. On the other hand, it is argued that the costs of the macroeconomic and financial turbulence that ensue when external deficits—private or public—become unfinanceable are frequently socialized, creating a public interest in ensuring that the private and the public sectors remain safely within their intertemporal budget constraints.

Second, instruments for affecting private savings and domestic investment are notoriously scarce, unreliable, and costly to use. Prudential regulation of the banking system, to discourage inappropriately risky lending by banks that benefit from implicit or explicit deposit insurance, may be called for, as are measures to prevent the increased domestic liquidity generated by rapid capital inflows from translating into excessively rapid growth in bank credit.

But the major instrument for affecting national savings is fiscal policy. Unfortunately, there is good theoretical and some empirical support for the idea that private savings may adjust to partly offset changes in taxes, particularly if these are perceived to be transitory. This means that large changes in tax policy may be required to offset swings in capital flows, which is not only politically and administratively difficult, but also inefficient in terms of public finance. Similar considerations apply to government spending. National budgets are more than just levers of macroeconomic policy; they are the chief means through which important social policies that determine the health, education, and national defense of the domestic population are realized. Subjecting these programs to radical fluctuations in response to movements in international capital flows may be necessary, but it is not costless.

Policy Issues

With this brief overview of the basic macroeconomics of international capital flows as a backdrop, we proceed to a discussion of policy issues raised by recent Latin American experience. This section focuses on a selected set of broad issues and policy strategies rather than on a comprehensive and detailed analysis of operational issues surrounding the policies toward international capital flows.

Prospects and Immediate Policy Response

The sharp reduction in capital flows to the region that occurred in the immediate aftermath of the Mexican crisis has eased. To several countries in the region, and many other emerging market economies outside it, capital inflows are once again large enough to pose the policy challenges that materialized in the 1990-1994 period. One reason for the renewed flows to the region is the substantial decline in U.S. interest rates that took place in early 1995. This development again highlights the region's exposure to fluctuations in world interest rates.

Despite the generalized recovery of capital flows to much of the region, flows to Argentina, Mexico, and Venezuela have remained sharply lower. Even there, though, are signs of recovery. For example, the government of Argentina has successfully reentered international financial markets, as have some Mexican borrowers, including Bancomex, Nafinsa, and the Government of Mexico, and a few private sector firms. However, in almost all cases, interest rate spreads have been fairly high, and loan durations short.

What are the lessons of this recent experience?

- First, the policy problem facing the region is not primarily managing a long- or medium-term scarcity of capital flows, but rather coping with their highly volatile nature. Large cross-border capital flows are likely to be an important feature of the international financial landscape for the foreseeable future, but they are likely to remain highly volatile.
- Second, the "tequila effect" of the Mexican crisis on several countries in the region in early 1995 shows that contagion effects are real: a country may experience shocks to the capital account as a result of developments in other borrowing countries, despite the fundamental soundness of its own policy framework. This creates vulnerabilities that require careful management. But the recent experience also shows that markets eventually make distinctions that reflect underlying fundamentals, and if a country can weather temporary financial storms with its macroeconomic policies and other fundamentals intact, international investors eventually respond accordingly.

The danger of volatile capital flows is that, in a fragile macroeconomic environment, even transitory capital account shocks can create disruptions large enough to validate the initially exaggerated fears about a country's creditworthiness. The challenge, therefore, is to construct policy regimes that create a robust and shock-proof economy capable of withstanding temporary shocks to international capital flows, while maximizing the developmental impact of international investment. We now consider five aspects of this challenge: exchange rate policy, debt management, fiscal policy, regulation of capital inflows, and the role of the international financial institutions.

Policy Regimes for a World of Volatile Capital Flows

Exchange Rate Policy

On this issue, no strong consensus has emerged from the recent experience. Advocates of fixed exchange rates note that the Mexican devaluation was followed by a major crisis. Supporters of exchange rate flexibility, on the other hand, point to the deep and growing macroeconomic costs of the monetary contraction generated by the classical balance of payments adjustment enforced by the Argentine convertibility plan. Clearly, neither fixed nor flexible exchange rates can eliminate the macroeconomic pain created by a sudden loss of international confidence. Can intelligent choice of exchange rate regime prevent such a loss of confidence? More generally, what are the lessons of recent Latin American experience for exchange rate management in a world of volatile capital flows?

First, recent experience highlights the importance of sustainability of a country's exchange rate regime. While there is room for dispute about the insulating properties of alternative exchange rate regimes, there is little doubt about the destabilizing consequences of being forced to abandon a regime that proved insufficiently robust to survive a large shock, from the capital account or elsewhere.[14] Thus, a case for the desirability of fixed exchange rates must rest upon a reasonable expectation that the regime will be robust enough to withstand the major shocks to which it will eventually be subjected.

In this context, two issues arise. The first is the need for a forceful fiscal response to protect a fixed exchange rate system after a sudden reduction in capital flows. Here, the recent experience of Argentina is revealing. The Argentine exchange rate system did not weather the 1995 shock automatically and painlessly. Protecting the system from collapse

---

[14] Hausmann, Gavin et al. (1995) provide evidence on both the insulating properties of alternative exchange rate regimes and the destabilizing consequences of unsustainability. Their estimates imply that Latin America has in the past paid a high price for switching between exchange rate regimes, which has often occurred because the existing regime proved to be unsustainable.

required a determined fiscal response—both tax increases and spending cuts—that were not only procyclical but also politically very costly. These actions were politically viable in Argentina because of the strong popular commitment to the convertibility law, grounded in the perception that the system is necessary to avoid the inflationary experiences of the past. In countries where public support for the exchange rate system is less strong, the fiscal response to capital account shocks may not be forthcoming, and the system is unlikely to survive a major shock.

The sustainability of fixed exchange rates also depends upon the strength of the domestic financial system. As the Argentine example indicates, the adjustment to a reduction of capital flows under fixed exchange rates involves a potentially sharp monetary contraction, which means both high interest rates and a cutback in credit extended to domestic borrowers. Under these circumstances, borrowers may experience difficulties in servicing their debts, and unless the banking system is robust, a highly disruptive banking crisis may emerge.

These considerations highlight the value of exchange rate flexibility in the aftermath of a shock to capital flows. In addition to the issues described above, there is the simple fact that capital account shocks require an adjustment of the real exchange rate, and it is macroeconomically less disruptive to achieve this through an adjustment of the nominal exchange rate than through changes in the myriad prices of the goods and services produced in a modern economy.

The main reason to forego the benefits of more flexible exchange rates is the desire to find a means of enforcing commitment by monetary policymakers to a stable and noninflationary policy, by reducing the scope for discretion in setting monetary policy. In some countries, the value of the commitment that is provided by fixed exchange rates may outweigh the loss of flexibility. But in other cases, rigid exchange rate commitments may be a very expensive way to impose discipline on monetary authorities. In such cases, is all lost in the struggle for discipline and commitment?

The answer is no. Advocates of rigid exchange rate regimes or unbreakable monetary rules as the only mechanisms available to solve the problem of time inconsistency in monetary policy may be unduly pessimistic. What matters is not that there be a single, unbreakable rule, but rather that policymakers communicate to the public a set of principles for policy formation that preclude opportunistic actions. But such principles need not preclude a policy response to changed circumstances, as long as the response is understood by the public to be part of a sensible and predictable regime.

Here, the experiences of Chile and Colombia are instructive. Neither country operates under freely floating exchange rates, and both countries have oriented monetary policy toward satisfying some form of exchange rate commitment. On the other hand, both countries have also retained substantial discretion in the management of their monetary and exchange rate policies. So, for example, when in 1994 and 1995 large capital inflows posed a difficult choice between maintaining their announced exchange rate bands and suffering inflationary pressures, or changing the bands to allow the exchange rate to appreciate, both countries chose to move the exchange rate bands. This policy adjustment was not perceived as a violation of policy commitments or a fundamental change in the rules of the game, but rather as a policy response to an external shock wholly consistent with the underlying policy regime. In contrast, the Mexican devaluation of December 1994 was viewed by many participants in international financial markets as a negation of previously made policy commitments, because in their view the authorities had made unconditional promises to maintain the previously existing exchange rate regime.

Thus, the conflict between commitment and flexibility in the choice of exchange rate regime can be alleviated by rules that are contingent, in reasonably predictable and well understood ways, upon unforeseen macroeconomic developments.[15] For such a regime to be credible, it must first be comprehensible, which means that the rules of the game need to be reasonably simple. But, as the examples of Chile and Colombia illustrate, they need not be so simplistic as to preclude an exchange rate response to a genuine and observable shock to the economy.

Debt Management

Debt maturity. The crisis that followed the Mexican devaluation was badly aggravated by the country's reliance upon short-term debt, which made the country highly dependent upon continual access to international credit markets to roll over much of the outstanding stock of debt. Even a temporary loss of market access placed the country in an untenable position, from which it was extracted only by deep and painful adjustment measures and by extraordinary efforts of the international community. Argentina was better able to weather the financial market turbulence of early 1995 because its longer-term debt profile relieved it of the need to access international financial markets during the worst of

---

[15] See Calvo (1992) for a more extended discussion of such contingent rules.

the international financial turbulence, and permitted the country to delay its return to those markets until after the panic had subsided and a strong adjustment package had helped reestablish its perceived creditworthiness. Put differently, the fact that investors did not have the opportunity to leave in herds in a short period of uncertainty gave authorities crucial time to react, adjusting fundamentals as required and restoring confidence.

Thus debt, and particularly short-term debt, can aggravate economic instability. Governments should therefore finance themselves with medium- and long-term debt to the maximum possible extent. And, to guard against potential negative effects of debt, it is important that the central bank hold a significant proportion of the country's expected debt service (including amortization) for the following quarters in highly liquid and readily available international reserves. This policy implies that if the debt is short run, it would need to have almost full backing in international reserves, over and above the coverage that is required to ensure prompt payment of imports and fluctuations in the demand for base money.

Denomination. If the stock of domestic debt is large and is denominated in domestic currency, it may be difficult to extend its maturity, given the inherent exchange risks involved and the volatility of inflation and interest rates. Also, it may be subject to self-fulfilling expectations of inflation. If investors believe that inflation will accelerate, they will demand a higher interest rate, which will aggravate the fiscal deficit and may force the government to accommodate the expected higher inflation. One means of reducing these problems is to denominate government debt contracts in a more stable unit of account, or to provide a mechanism to adjust returns, protecting them from the ravages of inflation. When debt is denominated in foreign currency we speak of dollarization; when it is indexed to some measure of domestic prices we speak of indexation. Choosing either of these two mechanisms would insulate interest rates from changes in the public's expectations of inflation or of exchange rate movements. Moreover, by protecting investors from changes in inflation, exchange rates, and domestic interest rates, these denominations may allow the government to extend the maturity of the debt.

However, if the government policy is unsustainable and will eventually require a major adjustment of the exchange rate or an acceleration of inflation, the stock of dollarized or indexed debt will be made effectively more expensive, precisely at the time when debt service is harder to maintain. Hence, denominating the debt in dollars or in an indexed unit should never be used as a means to postpone a needed adjustment. But if it is done in the context of a sustainable program, it may help to shelter that program against self-fulfilling expectations and to extend debt maturities.

Place of issue. Should public debt be issued domestically or abroad? This distinction is becoming less relevant with the liberalization of capital transactions, which implies that foreigners can buy debt issued domestically and residents can buy bonds issued abroad. Nevertheless, there are two important elements that should be taken into account. First, it is important to determine whether the debt is purchased by the domestic banking system, which is more likely to happen with domestically issued debt. This is so because banks purchase these instruments with resources obtained through short-term deposits. Moreover, they typically have guaranteed access to liquidity from the central bank in case of need. Usually, short-term domestic debt can be used for repurchase operations with the central bank. Hence, in practice, domestic debt held by banks is equivalent to interest-bearing money. Domestic debt is, therefore, more inflationary than debt held by foreigners, but is somewhat less sensitive to changes in international interest rates because it is demanded in part for liquidity purposes.

By contrast, demand for foreign debt may be more volatile, since it usually represents very small percentages of the holder's portfolio and these fractions may be very unstable. If it is short-term debt, this could pose very serious problems. Consequently, short-term foreign debt may not be as inflationary as domestic debt, but may be a dangerous source of volatility.

Fiscal Policy

Fluctuations in capital flows will, like other macroeconomic shocks, generally demand a fiscal response. The need for a swift response is particularly acute in the immediate aftermath of an adverse shock because such a shock often implies a sharp contraction in the availability of noninflationary financing of fiscal deficits. At the same time, its contractionary impact tends to increase fiscal deficits. But more generally, because capital account (and other) shocks are so much larger in Latin America than in industrial economies, the pace of fiscal adjustment needs to be more rapid; unfortunately, Latin America and the Caribbean cannot afford the protracted discussions and adjustments that may be tolerable in the industrial economies.

However, the political process through which fiscal policy is decided in the region is similar to that of most democracies. It typically involves an executive branch and a legislative branch. The executive has several spending ministries and the legislature may have two houses. Passage of a law requires discussion and vote in each house and reconciliation between the two versions. Going through the process takes time, and not only because of bureaucratic delay. As argued by Alesina and Drazen (1992), different constituencies may have incentives to delay adjustment, not because they believe adjustment is not needed but because, by obstructing a solution, they may be able to shift the burden of adjustment onto other groups.

Given the larger fiscal risks, the limited willingness of financial markets to finance the implied deficits with a prudent debt structure, and the inherent difficulty faced by democratic political systems in reacting quickly to budget problems, it is not surprising that fiscal policy has been such an important determinant of macroeconomic instability in Latin America. However, there are concrete policy strategies that can ameliorate the problem. Some strategies of particular relevance for the management of volatile international capital flows are to: (i) set precautionary fiscal targets, (ii) adopt budgetary rules and institutions that deliver quick responses, and (iii) institutionalize contingent rules for shock management.[16]

Set precautionary fiscal targets. In principle, it would be desirable to offset the contractionary impact of a sudden reduction in capital flows with a counter-cyclical fiscal expansion. The difficulty, of course, is that the shock also makes noninflationary financing for the implied budget deficits much more scarce, potentially creating the need for a procyclical fiscal contraction instead. This is particularly likely if the fiscal situation is precarious before the shock arrives. This creates the apparently paradoxical possibility that fiscal contraction is the appropriate response to sudden capital inflows and outflows, in the former case to limit the expansionary impact of the inflows, and in the latter case to bring the government's financing requirements into line with the new, lower availability of noninflationary financing.

If, however, countries adopt the practice of targeting a fiscal surplus during normal times, a counter-cyclical response to adverse capital account shocks becomes much easier to achieve. It is, after all, much easier

---

[16] The following discussion draws heavily upon Hausmann, Gavin, et al. (1995), which contains a much more extended discussion of these and related strategies for ensuring appropriate fiscal adjustment to shocks.

to permit a budget surplus to become smaller than it is to finance a budget that has moved into deficit in the aftermath of an adverse shock to the capital account. And, in the long run, the lower steady-state capital stock will reduce the likelihood that domestic and international investors will develop fears about the government's creditworthiness.

Adopt budgetary rules and institutions that deliver quick responses. The fiscal response to an economic shock is the outcome of decisionmakers working within a specific institutional context. This institutional context defines the terms of the budgetary debate, for example by establishing the relationship between the Ministry of Finance and the spending ministries and between the executive and the legislature, and thus helps determine budgetary outcomes. Alesina, Hausmann, Hommes, and Stein (1995) provide evidence, for example, that the budgetary rules that exist in different Latin American economies have had important effects on long-run fiscal outcomes. Similarly, some institutional features of budgetary management are conducive to rapid and effective fiscal response to shocks, while others increase the risk of gridlock and delay. For example, it is commonly the case that the executive proposes a budget that must be acted upon, after debate, by the legislature. If there is no deadline, incentives for the legislature to come to a timely agreement may be insufficient to prevent deadlock. On the other hand, if there is a deadline after which, for example, the executive's proposed budget comes into force, the legislature is provided with much stronger incentives to enact budgetary legislation promptly.

Institutionalize contingent rules for shock management. Explicit rules that specify the fiscal response to major economic contingencies can also promote effective fiscal adjustment. Stabilization funds, such as for copper in Chile and for oil and coffee in Colombia, are one form of automatic spending rule that is particularly well suited for clearly identified sources of revenue volatility. But such automatic adjustments can also apply to revenues as well. For example, in Ecuador the contingent rule specifies that if oil tax revenue falls below the budgeted level, the domestic price of gasoline must be increased to make up for the fiscal shortfall. Sharp fluctuations in capital flows will, through their effect on international trade, domestic spending, and other determinants of the tax base, exert indirect but potentially powerful influence on fiscal outcomes. It would be useful to reach agreement, before such shocks arrive, on rules that determine how to absorb the fiscal implications.

Regulation of Capital Inflows

Integration into world financial markets holds enormous promise for the economic development of Latin America. But the policy problems created by volatile capital flows are serious as well, which raises the question whether some regulation of such flows is called for, and if so, how it should be structured. There is no consensus on this issue, but it seems safe to say that the turmoil that afflicted international financial markets in early 1995 has increased the respectability of proposals to regulate international capital flows, not as a panacea for the problem, but as one of several policy instruments that can be enlisted to reduce the volatility of capital flows and ameliorate their macroeconomic consequences.

A first question is whether such regulation can be effective at all. Experience shows that the effectiveness of restrictions on financial flows tends to diminish gradually over time, as market participants find ways around the restrictions, except perhaps in very highly regulated financial systems. If regulation of capital flows becomes totally ineffective it will be, at best, a useless instrument, and it may be worse than useless if the actions taken to circumvent the restrictions are themselves economically costly. The potential effectiveness of restrictions on capital flows is an empirical issue that needs to be settled on a country-by-country basis. It should be noted, however, that permanent and total isolation from fluctuations in international capital flows is not a plausible or sensible goal, and that the difficulties of enforcing the regulations required to achieve this draconian end make them infeasible.

What is at issue is the scope for altering, at the margin and perhaps temporarily, the magnitude and types of flows that a country experiences. The cases of Colombia and Chile are revealing here. Both countries have made substantial efforts to regulate capital inflows, particularly of short-term debt. Both have nevertheless faced significant policy challenges from high rates of capital inflow, suggesting that the restrictions were not a panacea. But both countries were also essentially unaffected by the tequila effect that gripped much of the continent in the aftermath of the Mexican devaluation. This suggests that regulation of capital inflows need not be perfectly effective to provide meaningful protection against international financial turmoil.

A second question is whether inflows and outflows should be treated symmetrically. On the one hand, it might be argued that the policy problem stems not from inflows themselves, but rather from the possibility of sudden outflows, so that attempting to regulate the outflows would be a more direct approach to the problem. However, regulating capital outflows may be much less effective than regulating inflows. To paraphrase an old saying about the stock market, inflows are generated by investors' hopes and outflows by their fears, and the reality is that fear is a more potent motivator than is hope. More to the point, there is a danger that restrictions on capital inflows will be used, as they have been in the past, to insulate misguided economic policies from market discipline. Even if this is not the purpose or the result of the policy, it may lead to difficulties by creating an unfortunate perception.

These problems apply less forcefully to regulation of capital inflows. For that purpose, another macroeconomic instrument must be found to cope with the macroeconomic boom resulting from inflows that derive from international confidence in economic prospects and the policy regime, and to ensure that the inflows do not make the economy excessively vulnerable to sudden withdrawals of capital. Such instruments are therefore unlikely to be, or be perceived as, substitutes for policy discipline. And because capital inflows are motivated by favorable interest rate differentials, which are substantially smaller than the losses that investors may fear when attempting to withdraw their capital, regulation of inflows is more likely to be effective.

A third question is whether different forms of capital inflows should be treated differently. On the one hand, the short-term macroeconomic impact of capital inflows is largely independent of the financial instrument through which they are effected. However, different kinds of capital inflows expose the economy differently to the risk of sudden, subsequent withdrawals of capital. In particular, the danger posed by the rollover risk that was highlighted in our discussion of public debt management applies equally to privately issued international debt. There may therefore be a special policy interest in ensuring that short-term international debt is issued prudently by private as well as public borrowers.

The Role of International Monetary Coordination

The greatly increased international mobility of capital has increased the magnitude of the financial problems with which policymakers must cope—both directly, by increasing the magnitude of the financial shocks that affect a given economy, and by virtue of contagion effects through which a financial crisis in one country can be transmitted to other vulnerable economies. But the amounts of capital that can be raised through normal channels are woefully inadequate to ensure orderly adjustment to the massive balance of payments shocks that occur in today's integrated financial markets.

The international rescue package created to cope with the financial disorder that followed the Mexican devaluation highlights the inadequacy of current arrangements for international monetary cooperation in a world of highly mobile capital. The rescue effort took many weeks to mount, required highly unorthodox methods on the part of both the International Monetary Fund and the U.S. Treasury, and remains controversial enough to create strong doubts that it could be replicated for any other country. This raises the question whether and how existing mechanisms of international monetary cooperation need to be reformed.

If the international community is to cope with these larger international financial shocks, it will need access to larger financial resources. The main drawback of making such resources available for rescue operations is the moral hazard created by investors' and policymakers' perceptions that they will be bailed out in the event of a crisis. This may create incentives for domestic policymakers and investors to behave in an excessively risky manner. One response to this possibility is to create early warning systems, which would improve international policymakers' capacity for predicting and heading off future crises. The implied package of financial support to resolve crises, with surveillance to head them off, is nothing new, but rather an expansion and intensification of the International Monetary Fund's main lines of business.

While monitoring macroeconomic developments and responding effectively to crises as they arise is clearly important, more can be done to reduce countries' vulnerability to capital account shocks and increase their capacity to deal with them. As we have tried to convey in this paper, both an economy's vulnerability and its capacity to respond effectively depend upon key structural characteristics. These include, among other factors: the depth and robustness of the domestic financial system, the suitability of labor market regulations, the nature of budgetary institutions, and the capacity of the domestic political system to forge a lasting consensus on economic policy. Short-term macroeconomic management cannot substitute for systemic reforms in these areas. While such reforms are, in the final analysis, the responsibility of a country's own policymakers and populations, international agencies can provide financial assistance and, no less important, contribute the expertise acquired through policy research and first-hand experience with similar reforms in other countries.

Michael Gavin is Lead Research Economist and Ricardo Hausmann is Chief Economist at the Inter-American Development Bank. Leonardo Leiderman is Professor of Economics at Tel-Aviv University.

References

Alesina, Alberto and Allan Drazen. 1992. "Why Are Stabilizations Delayed?" *American Economic Review* (December).

Alesina, Alberto, Ricardo Hausmann, Rudolf Hommes, and Ernesto Stein. 1995. "Budget Institutions and Fiscal Performance in Latin America." OCE Working Paper. Inter-American Development Bank.

Blanchard, Olivier and Nobu Kiyotaki. 1987. "Monopolistic Competition and the Effects of Aggregate Demand." *American Economic Review.*

Calvo, Sara and Carmen Reinhart. 1994. "Capital Flows to Latin America: Is There Evidence of Contagion Effects?" mimeo. International Monetary Fund.

Calvo, Guillermo, Leonardo Leiderman, and Carmen Reinhart. 1993. "Capital Flows to Latin America: The Role of External Factors," IMF Staff Papers.

Chuhan, Punam, Stijn Claessens, and Nlandu Mamingi. 1993. "Equity and Bond Flows to Latin America and Asia: The Role of External and Domestic Factors." mimeo. World Bank.

Cooper, Russell and Andrew John. 1988. "Coordinating Coordination Failures in Keynesian Models." *Quarterly Journal of Economics* (August).

Fernandez-Arias, Eduardo. 1993. "The New Wave of Private Capital Flows: Push or Pull?" mimeo. World Bank.

Gavin, Michael. 1993. "Unemployment and the Economics of Gradualist Policy Reform." mimeo. Columbia University.

Hausmann, Ricardo, Michael Gavin, et al. 1995. "Overcoming Volatility in Latin America," in Inter-American Development Bank, *Economic and Social Progress in Latin America, 1995 Report.* Johns Hopkins University Press, Washington, D.C.

Mussa, Michael. 1982. "Government Policy and the Adjustment Process," in J. Bhagwati (ed.), *Import Competition and Response.* University of Chicago Press, Chicago.

COMMENTARY TO PART I

Michael Bruno

I have spoken elsewhere of global optimism versus local pessimism, and indeed, in the wake of recent events it is important not to be overly pessimistic about the prospects for capital flows to developing countries. There are good reasons to be optimistic from a global perspective. What underlies the recent surge in capital flows to developing countries is, after all, a stock adjustment of the portfolios of savers in industrial countries. The demographics of these countries indicate that their saving rates are likely to continue to rise. At the same time, the diversification of portfolios should continue, especially since profits are likely to be higher in reforming countries than in these investors' home countries.

In 1913, 46 percent of the United Kingdom's foreign equity portfolio was invested in what are today's emerging markets, and of that, 20 percent was invested in Latin American countries.[1] Present levels of industrial country investment in the developing world are far lower. The typical U.S. pension fund, for example, has no more than 1 to 2 percent of its assets invested in emerging markets. We might never see that share return to the values at the turn of the century, but even if one assumes that it will increase to no more than 5 percent over the next ten years, that implies a robust and sustained rate of growth of gross capital flows into developing economies. Of course, gross flows are not net flows. The latter depend in part on what happens to current accounts. The danger is not that capital flows will be lacking; rather, the danger is that current account deficits will be too large. Indeed, they already are too large in some Latin American countries.

The composition of flows is also very important. In recent years, some 40 percent of global capital flows to developing countries took the form of foreign direct investment. But that proportion has varied greatly from region to region. In both Latin America and East Asia, private capital inflows in recent years have amounted to some 3 to 4 percent of GDP. But, as Larry Summers notes, whether this is a good thing or not de-

---

[1] Note also that in the period 1885-95, Great Britain ran a current account surplus of 49 percent of GDP and that a large share of that surplus was invested abroad, much of it in the countries we now call emerging markets. See A. Kenwood and A. Lougheed, *The Growth of the International Economy, 1820-1990*, Routledge (1992). See also Lant Pritchett, *Capital Flows: Five Stylized Facts for the 1990s*, World Bank (1995).

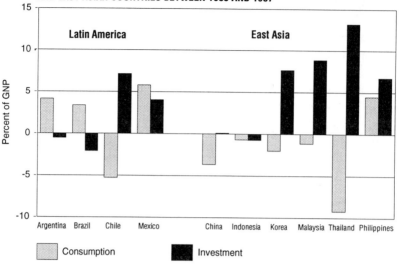

CHANGE IN CONSUMPTION AND INVESTMENT IN SELECTED LATIN AMERICAN AND EAST ASIAN COUNTRIES BETWEEN 1985 AND 1987

pends on whether those inflows go to finance investment or consumption. On this score the two regions differ: capital flows to East Asia (except the Philippines) have primarily financed private investment, whereas in Latin America as a whole (but not in Chile) they have financed an increase in consumption (see figure above). Within Latin America, Mexico has probably done the least well on this criterion, but Brazil and Argentina have also run current account deficits largely to finance consumption rather than investment.

The two regions differ on a number of other measures as well. The share of exports in GDP has been higher on average in East Asia than in Latin America, at about 11 percent compared with 7 percent. (Chile again looks more like an East Asian country in this respect.) The ratio of external debt to GDP has averaged about 130 percent in the Asian countries versus 253 percent in Latin America, and the maturity structure of the debt also differs. The more stable financial environment of East Asia has favored longer maturities, whereas in Latin America, in extreme cases such as the recent Mexican episode, a large stock of the debt is held in what are essentially overnight markets. Most important, the regions differ in their external competitiveness and in the priority given to maintaining and enhancing it. Most of the East Asian countries have absorbed large capital inflows without a real appreciation of their currencies. Moreover, the proportion of foreign direct investment in total inflows has generally been higher and taken the form of new, greenfield investment rather than privatization of state enterprises. This has not been the case in Latin America. The currencies of several countries in the region have experienced large real appreciations, which have caused a significant loss of competitiveness. Foreign direct investment in the region has been directed largely to privatization. This is undoubtedly good, and it has helped to reduce fiscal imbalances. However, it might not be associated with the same gain in productivity as greenfield investment.[2]

The East Asian countries have generally shown greater fiscal restraint. Although the Latin American countries have considerably improved their fiscal performance, with most achieving successful stabilization, the picture is somewhat less flattering if one looks not just at central government finances but at the consolidated public sector deficit as well. In some cases, there are hidden deficits at the provincial or the local level. Development banks have often been used as instruments for directing credit to nonviable but politically protected projects that eventually turn sour, leaving the banks with a large portfolio of nonperforming assets. Certainly, the comparison across countries shows how important fiscal restraint is to the overall economic outcome.

Can structural reforms succeed in guiding capital flows into investment in export industries rather than into consumption? We know that it is not easy to increase private saving rates. But one way to do it, which has been successful in Chile, is to introduce mandatory saving programs, for example through pension reform. I think some other Latin American countries are also moving in this direction, and this is encouraging. One could also use fiscal incentives to increase corporate savings, but without an otherwise undistorted tax system, these measures can miscarry.

Finally, there is the need for reform of the institutional and regulatory framework governing banking and the capital markets, which are relatively thin in the countries in which this strong capital impulse is being felt. At the global level, I have not made up my mind whether it would be a good thing to establish a large and institutionalized lender of last resort, as, for example, in some current proposals to establish a rescue facility at the IMF. One can offer good reasons for such a policy, but the familiar moral hazard problem is not to be dismissed lightly. It is not so much a question of having or not having fire engines as of where the

---

[2] Interestingly, in 1994 foreign direct investment in Latin America continued to increase, despite a decline in other forms of capital flows. This might have been due to the continuing trend toward privatization.

fire engines should be kept, and when and by whom they should be dispatched. Should funds be kept in the coffers of central banks rather than with the IMF? There are too many unanswered questions. I fully agree, however, on the need to improve early warning systems: countries should fully disclose their contingent liabilities and how their public debt is being financed. Full transparency is needed so that surveillance bodies such as the IMF will have all the information they need in real time, if for no other reason than to be well prepared to decide what their policy should be when another crisis happens.

Michael Bruno is Vice President, Development Economics and Chief Economist at the World Bank.

Domingo Cavallo

My conclusions rely entirely on the Argentine experience and therefore are based on a sample of one. Yet the Argentine experience is relevant to the rest of Latin America, first of all because of our history. Over a period of six decades, from the 1930s to the 1990s, Argentina has tried everything: fixed exchange rates and flexible exchange rates, capital controls and complete freedom of capital flows, military governments and democratic governments, Peronism and conservatism. Of course there have been some constants in the Argentine economy over those six decades: massive government spending; increasing government intervention in the economy, which encouraged corruption and inefficiency; and an ever more complex tax system in which opportunities for tax evasion proliferated. But probably the most striking change over those six decades was that the Argentine economy became less and less transparent, not just to foreign investors, foreign bankers, and the multilateral agencies, but to the Argentine people themselves. It became increasingly difficult for citizens, and for politicians, to understand what was going on in Argentina. That loss of transparency explains why Argentina suffered several decades of stagnation, punctuated by deep recessions, with increasing inflation and finally hyperinflation. Fortunately, however, the Argentine people and their leaders learned the lessons of their six long decades of suffering, and the lessons signified a return to common sense.

The capital inflows of the last four or five years were very beneficial for Argentina. Not only did they help increase overall investment, but also, and more important, they helped to make the new investment much more productive, thereby increasing the rate of growth of overall productivity. Foreign investment—including direct investment, portfolio investment, and trade financing—amounted to 3 percent of GDP in 1993 and 1994. Inflows were somewhat less than that in 1991 and 1992, but in total over the four years from 1991 to 1994, Argentina received the equivalent of 7 or 8 percent of one year's GDP. Meanwhile the Argentine economy grew by 7.7 percent a year, for a total of 34 percent in those four years. Total factor growth in those years was only 3 percent a year; the rest of the expansion was therefore due to productivity growth, which increased by almost 5 percent a year. Investment in the Argentine economy—financed both by domestic saving (which rose from 14 percent to 17 percent of GDP between 1990 and 1994) and through foreign saving—was clearly very productive. The source of that rise in productivity was a thorough deregulation of the economy: Argentina not only opened its economy to foreign investment and foreign trade, but at the same time deregulated most of the domestic economy, to allow markets instead of the government to allocate resources. In addition, we privatized most of our state enterprises, concentrating the activities of government on setting macroeconomic policy and supplying those basic services that only government can provide.

We also adopted some unique institutions that helped to restore common sense in the way that not only ordinary Argentines, but also our politicians, think and behave. Our monetary policy, for example, is based on a very simple arrangement: our central bank was in effect transformed into a currency board. A currency board may print only as much money as it has reserves of foreign exchange; thus, all the money that circulates in the Argentine economy is backed by foreign exchange. Such an arrangement sends the message that the monetary system cannot create credit simply by printing money, and that one cannot increase the competitiveness of the country by devaluing the currency. Once this message sinks in, all economic actors—politicians as well as private businesspeople—will come to understand that governments that run deficits, and enterprises that find themselves in financial trouble, will obtain financing only if someone can be found to lend to them on a voluntary basis. No longer can the government depend on the central bank to finance its deficits, because the central bank cannot create money to lend. No longer can illiquid or insolvent financial institutions count on the central bank to print money to cover their debts. Eventually, all would-be borrowers will begin to comprehend that to obtain credit, they must create confidence: enterprises must show that they have projects that are worth investing in; governments and enterprises must show that they are capable of managing their affairs efficiently.

By abandoning the option of devaluation, the currency board arrangement signals to enterprises what is, after all, a common sense truth: that the goods and services they produce must be competitive both in the local market and abroad. A country with a currency board cannot reduce wages across the board through devaluation; the only way for producers to increase their international competitiveness is to become more productive, improving the ways in which they use their human and capital resources and reducing the cost of production and the cost of investment by adopting the best technologies. Politicians, too, will begin to understand that the best way to help the private sector become competitive is to remove distortive taxes and regulations that increase the cost of production. All participants in the economy thus receive a valuable reeducation in how an economy works, or should work.

This reeducation has allowed Argentina, when capital once again began to flow in from abroad, to use that capital efficiently to bring about rapid growth—even as the economy was disinflating. Argentina is perhaps the only large Latin American economy in which annual inflation is below 4 percent. Chile and Colombia are often presented as examples of successful stabilization, yet inflation there remains in or near the double digit range. In my view, real economic stability means getting inflation close to zero—to rates equivalent to those in Europe, North America, or Japan. That is what we are trying to do, and have done, in Argentina. Toward that end, our peculiar monetary arrangement was essential, but it was helped by two other critical changes. One was to give the Argentine people the freedom to choose which currency to use in writing contracts and conducting everyday transactions. The other was to eliminate the practice of indexing for inflation. This again is merely common sense economics: indexation only reduces the transparency of economic decisions and complicates all contracts and transactions to which it is applied.

Few would dispute that the capital inflows of the early 1990s helped the Argentine economy. But I would argue, more controversially, that the capital outflows that Argentina has experienced more recently have helped, too. They helped because, in spite of the Argentine economy's impressive progress toward transparency during the last few years, some politicians still did not get the message. They did not understand, for example, that it was necessary to break the link between the provincial banks and the provincial governments, because to the extent that provincial governments can borrow from their provincial bank, they are not subject to fiscal discipline. Some politicians also refused to accept the reality that one cannot continue to raise retirees' benefits by printing money. Higher pensions might seem no more than social justice, but even more unjust are the effects of inflation. Thanks to the pressures exerted by the recent outflows, several important reforms that the executive branch had proposed to the Congress year after year without success have at last been approved.

The recent crisis also demonstrated the need in Argentina for an institution that can act as lender of last resort to the financial system. We are in the process of creating such an institution, but it will not be one that accomplishes that task by printing money and reintroducing inflation into the economy. Instead, it will obtain its funding either from fiscal resources or from willing lenders. Some will argue that a currency board arrangement deprives the central bank of any possibility of acting as lender of last resort. This is not so. A currency board can serve as a lender of last resort; what it cannot do is create the money that it lends ad libitum, whether for bailing out banks or for financing a fiscal deficit.

For these reasons, we in Argentina believe that the recent crisis will help us take another step toward improving our economic organization. If we still need to learn one lesson, it is that much remains to be done to make our institutions truly transparent from the perspective of portfolio investors. In particular, we need to communicate more effectively to foreign investors what is happening in our country. Restoring credibility in the Argentine economy is thus a matter of educating not only our people and our politicians, but prospective investors as well.

Domingo Cavallo is the Minister of Finance and Economy and Public Works of Argentina.

Guillermo Perry Rubio

The world economy is entering an expansionary phase, with higher interest rates and diminishing international liquidity. As a result, capital flows from industrial countries to the developing world will be reduced during the second half of this decade. This will represent a significant change in the economic environment, and Latin American governments must be extremely cautious to smooth the impact of external financial shocks on their domestic markets.

Latin American countries must adjust their macroeconomic balances to the new financial environment. As the paper points out, the current macroeconomic imbalances in the region differ from the much larger disequilibria of the first half of the 1980s. Whereas in the earlier period most Latin American countries had large budget deficits, and their current account deficits were primarily financed with external public debt, in this decade the most prominent source of macroeconomic disequilibrium has been the saving/investment gap in the private sector, which has increased current account deficits. Countries with a strong dependence on foreign capital must adjust by lowering their current account deficits. This will imply a real depreciation of their currencies, and—to support current investment levels and sustain stable, dynamic growth—an increase in their saving rates.

Nevertheless, the capital outflow that followed Mexico's devaluation shows a certain degree of overreaction to the shocks of late 1994. As the paper explains, capital markets are imperfect and do not provide insurance against the risks associated with fluctuations in the magnitude of these flows. The contagion effect is also important in explaining how the region was affected by the Mexican crisis.

Colombia is prepared for the new economic environment. The new government that took office in August 1994 has set the objectives of maintaining a stable real exchange rate and avoiding dependence on short-term capital flows. Macroeconomic equilibrium and a fair degree of control over short-term flows are the principal means of attaining those objectives. In 1994 Colombia had a current account deficit of around 4.1 percent of GDP, due in part to exceptionally high investment in developing new oil fields and pipelines; the country simultaneously enjoyed a budget surplus, which will be maintained in the coming years. The current account deficit will decrease after 1995 as the new oil fields bring significant increases in export revenues. A portion of both oil and coffee export revenues will be saved abroad, with the objectives of increasing fiscal discipline and the domestic saving rate, and reducing the pressures for currency appreciation. The smaller projected current account deficits will be financed mainly through foreign direct investment (in oil, manufactures, and infrastructure projects) and long-term debt. Colombia's debt/GDP and debt/export ratios will diminish in the years ahead, as they have over the last decade.

The volatility of capital inflows has led many countries, including Colombia, to use currency bands, within which the exchange rate may fluctuate, rather than fixed or completely flexible exchange rates. Some flexibility in exchange rates helps to absorb the effects of capital movements and allows the domestic authorities to exercise some degree of monetary control. On the other hand, a currency band contributes to stabilizing the exchange rate, moderating expectations, and minimizing speculative currency attacks.

Without capital controls, attempts to sterilize the monetary effects of capital flows are self-defeating and impose high costs on the monetary authorities. Chile and Colombia, as the paper points out, adopted a policy of partial sterilization of capital inflows, which helped isolate their economies from external shocks. Together with a mix of flexible exchange rate policy with currency exchange bands and some degree of capital controls, sterilization makes it easier to distribute the effects of capital inflows or outflows among exchange rate movements, changes in domestic interest rates, and financial costs assumed by the central bank. In such a context, sterilization is a viable policy, but only if the country has the capacity to regulate capital flows.

As an economy becomes more integrated into the world economy, the capacity of regulators to enforce capital controls is reduced. However, we in Colombia are strongly convinced that the authorities still have an important role to play in reducing short-term capital flows.

Colombian economic policy has long sought to control short-term capital flows by imposing restrictions on short-term debt and taxing speculative capital inflows. The new government, in coordination with the independent central bank, introduced additional measures in this direction by mid 1994. Although we are aware of the limits to any kind of administrative controls, we have achieved satisfactory results with such a policy. Together with macroeconomic stability, capital controls helped to isolate the economy from the negative effects of the large financial shocks at the end of 1994.

An important lesson of Colombia's long history of capital controls and of the recent Chilean experience is that the simpler and more general the controls are, the easier it is to implement them. Colombia's current rules are quite simple: credits with maturities of less than five years have to constitute a high deposit at the central bank. The only exceptions are for trade financing with a term of less than four months, financing of imports of capital goods, and export credits channeled by the government-owned Colombian export bank (Bancoldex).

We have also been cautious in authorizing portfolio investment in Colombia. International investors have ample investment opportunities in the stock markets, but only recognized international mutual funds are allowed to participate. The objective of these measures is to attract institutional investors who will make investment decisions based on an evaluation of the long-term prospects of the Colombian economy.

Short-term portfolio investment must be treated differently from long-term and foreign direct investment (FDI). It is well known that FDI brings stability and positive externalities in developing economies. Technology transfer and management skills are among the most important contributions of foreign companies to our economies. In Colombia, FDI flows have increased substantially since the liberalization of the economy and the introduction of new policies aimed at attracting foreign capital: FDI inflows are expected to reach 3.6 percent of GDP in 1995, up from 2.4 percent in 1994 and 1.4 percent in 1993. The government expects that foreign capital will play an important role in manufacturing, oil, mining, and infrastructure-related projects in the years to come.

Fiscal policy is a very important tool for counteracting medium- or long-term cycles in capital flows. However, in the short term it is extremely inefficient to use fiscal policy as the only response to changes in international financial markets. As the paper explains, there has already been a significant convergence toward fiscal equilibrium in Latin America during the 1990s. Hence fiscal policy will not play as crucial a role as it did in the austerity programs after the debt crisis. Since most of the current account deficits that countries in the region have experienced in the 1990s are explained by greater liquidity and external financing to the private sector, there will be an endogenous reduction in private sector absorption to attain macroeconomic equilibrium. An important contribution of the authorities during the adjustment process will be to maintain a strict and orthodox monetary policy.

It is clear that international support is necessary to resolve balance of payments crises. We have learned that crises precipitated by sudden capital outflows are more explosive than those related to current account difficulties—and harder to resolve. A more rapid response mechanism and an increase in the resources devoted to resolving balance of payments difficulties are needed, but at present the international monetary system is not prepared to handle these problems. It is important to distinguish the objectives of the International Monetary Fund—the institution primarily responsible for helping economies that find themselves in balance of payments crises—from those of other institutions such as the IDB or the World Bank, whose job is to provide long-term financing. The IMF should be strengthened and reformed to provide it with the capacity to act as lender of last resort at a global level.

Guillermo Perry Rubio is Minister of Finance and Public Credit of Colombia.

Lawrence H. Summers

The problem of how to approach capital flows provokes a range of differing opinions. Nonetheless, it is encouraging to recognize that the entire discussion of developing economies takes place within an area of consensus that is far larger than would have been the case even a decade ago. No one who discusses emerging markets today can doubt the fundamental importance of stabilization. No one can doubt the fundamental importance of openness in promoting economic growth, or the role played by the private sector as a critical engine of such growth. It is a sign of progress that our discussion today is much narrower than the discussion we would have had ten or fifteen years ago. The International Monetary Fund and the multilateral development banks have had a great deal to do with that progress.

I would like to propose some answers—very tentative answers, to be sure—to three questions that I think must be addressed in any discussion of how countries should treat capital flows. First, when are capital inflows a problem? Second, what can countries facing strong capital inflows, or excessive volatility, do in response? And third, how can the international system respond?

First, when are capital inflows a problem? The former Chancellor of the British Exchequer, Nigel Lawson, attempted to provide an answer to that question. He proposed that large flows are not a cause for concern if they can be attributed largely to private sector requirements and are not due to a public sector deficit. I agree with his formulation to the extent that the absence of a public deficit is a necessary condition for confidence that capital inflows will not create difficulties. But I do not believe that the lack of a public deficit alone is a sufficient condition for confidence.

I would suggest three critical tests to determine when a potential problem exists. First, policymakers and analysts must look at how capital flows are being used. For nations as well as for people, borrowing to finance investment is seen as healthy, but borrowing to finance consumption is much more problematic. When the lion's share of inflows is being used for investment, there is the presumption that the economy is generating the capital to repay these obligations. That presumption is much safer when investment is taking place largely in the export sector rather than in the nontradable goods sector. Similarly, the national saving rate can offer some evidence as to whether capital inflows are being used to finance unsustainable consumption or investment. The national saving rate also provides an indicator of a nation's ability to weather a sudden diminution or reversal of capital flows. It is a basic truth that Latin America's lower saving rate, in relation to Asia's, explains the former's poorer economic record and weaker resistance to capital flow volatility.

Second, analysts must examine the terms under which capital is being lent or invested. When the money is entering on terms that are steadily more favorable and more secure for the borrower, there is much less need for concern. When flows are proceeding on steadily less secure terms for the borrower and more secure terms for the lender, greater concern is in order.

How can one determine whether terms are improving or deteriorating for the borrower? The maturity of outstanding debt is one indicator. There is less need for worry when maturities are lengthening, and more when they are shortening. Similarly, when debt is increasingly denominated in local currency, there is much less of a problem than when it is increasingly denominated in foreign currency. Last, when flows are going increasingly into direct investments or into portfolio or equity investments rather than debt investments, there is much less of a problem. On the other hand, when the pattern is one of increasingly short-term, increasingly foreign currency-denominated debt, that would suggest that a serious problem is arising.

The third criterion for determining when a problem exists centers on the size of capital flows. It is unlikely that any country can, over a long period of time, borrow more than 5 percent of its GNP annually unless it is growing at a very rapid rate. The adjustment required simply when the rate of growth of debt suddenly has to be reduced can be very dangerous. That is the case even when the total stock of debt or investment is not falling and people are holding the stock that they want to hold. For that reason, excessive growth of outstanding debt, or, equivalently, an excessive current account deficit, is dangerous.

I think that analysts and policymakers seeking to determine when countries must adjust to potential capital flow problems can benefit from adhering to three criteria: investment versus consumption, the terms under which capital inflows are coming, and the quantity of those flows.

The second question that policymakers must answer is, How should countries respond when they judge that capital flows are a problem? I think the one point on which almost all would agree is that policymakers should respond conservatively. That suggests a number of specific approaches.

First, there is a natural human tendency to suppose that periods of substantial inflows represent a permanent change, whereas periods of capital outflows represent a transitory disturbance. I think that experience suggests the opposite is true. As a rule, inflows should be treated as temporary, and outflow pressures as permanent. The policy responses are likely to be much healthier if that view is internalized.

The second general rule must be that responses are less painful if implemented sooner and more quickly. The same policy response taken preemptively will generate a much greater response, and a much greater corrective effect, than if taken under duress.

That rule calls forth three further rules. First, countries must respond through fiscal policy. For most countries, an improved budgetary position is the right response to pressure from both capital inflows and capital outflows. For the vast majority of nations, a larger budget surplus or a smaller budget deficit would represent an important contribution to economic health. If economics and accounting have one thing to teach politicians, it is that budget deficits are not an alternative to tax increases or reductions in expenditures. They are merely a way of postponing them, and one that carries substantial risks.

Second, countries should be very cautious about sterilizing changes in reserves, particularly when they are associated with capital outflows. That conclusion results from the principle suggested above, that one should treat outflows as likely to be permanent, and inflows as likely to be temporary.

The final point regarding how countries should respond to changing or volatile financial circumstances is that they should respond aggressively, with changes in the tools and methods of financial supervision and regulations. In my judgment, it is clear that we would all rather live in countries that capital is trying to get into than in countries from which capital is trying to get out. That suggests that countries should be very cautious about imposing capital controls with the objective of discouraging capital inflows. At the same time, new financial techniques present challenges to stability. It is appropriate for supervisory authorities to think about reserve requirements and new regulations, and to be prepared to respond aggressively to changes in the pattern of capital inflows, through improvements in regulation and supervision.

Let me turn to the final question that policymakers must address, namely, How can the international community respond to this new environment of vastly greater and more volatile capital flows to emerging and other markets? First, I think it is important to recognize that, despite the enormous changes we have seen in international capital markets, the basic tenets still hold. Money borrowed and used wisely serves nations well. But money borrowed and used poorly throws nations into difficulty.

In evaluating the international financial system and how it must adapt, I would highlight four issues. The first is the overwhelming importance of transparency. The need for financial transparency is something of a cliché, so I don't think it is treated with the importance that it deserves. Timely and frequent publication of comprehensive data on national accounts, on monetary conditions, and on central bank balance sheets makes an important contribution to international confidence in those countries that practice it. It signals that problems may arise and calls market and policy responses into play at an early stage, when those responses are most effective. Transparency therefore exercises important discipline on policymakers, so that they cannot slip and slide through a difficult situation.

I believe it is very important that the world move quickly to substantially raise international norms and standards for the publication of financial information. This is a task in which the international financial institutions have a role to play.

Second, we have to improve our tools and our means of surveillance. Policymakers and analysts must systematically examine the criteria I have suggested for judging capital inflows. In part, this is a task for the international financial institutions. But in very large part, it is also a task for the markets. Increased emphasis on transparency is something the markets should insist upon.

Third, the official sector needs to be able to respond in extraordinary circumstances when problems of confidence arise. The dictum that there is a need for lending of last resort in situations where there are liquidity problems but not solvency problems has a role in the international arena.

Fourth, there may be a need for enhanced coordination of debt restructuring, involving both private and public creditors, when nations do run into difficulties. It is sometimes in the interest of all creditors to coordinate their response to debtors' difficulties, while insisting on conditions that will ensure a return to economic stability. The fact that sovereign debt is now widely traded means that the old methods of coordinating bank debt restructuring may no longer be applicable.

It is right to worry about moral hazard. Moral hazard is a very serious problem. But I do not side with those who believe in abolishing the fire department because they think it will encourage people to smoke in bed. In just the same way, it is appropriate for the international community to respond at times of crisis, particularly when the crisis is of a systemic nature, as in the case of Mexico. The leaders of the Group of Seven nations gave careful consideration at their economic summit in

Halifax to how we can begin to consider modifications to the international financial architecture. We will have to give careful thought to how to maintain the capacity to respond at moments of extraordinary duress. But any solutions must ultimately rely on what is most important to international financial stability, namely, market incentives and strong policy decisions by particular countries. These are the ultimate guarantors of stability and confidence.

Lawrence H. Summers is Undersecretary of the U.S. Treasury.

CONCLUSION TO PART I

Jacob Frenkel

The comments reflect a significant convergence of macroeconomic perspectives and agreement on the proper policy response to imbalances. We now operate within a global system, not just of international transactions but of intellectual concepts as well. Economists and policymakers now share a common language, a common professional vocabulary. The world has indeed become a global village.

The nature of economic shocks, particularly the interaction of domestic and external shocks, has been the focus of the discussion. On the question of the proper fiscal response to such shocks, the commentators gave a rather surprising and conservative answer: whatever the nature of the shock, policymakers should always bring the budget back into balance. In a similar vein, Domingo Cavallo said that capital flows in either direction—outflows as well as inflows—are good for a country. The bottom line, however, is that even when a well-designed financial system is built with two-way streets, so that traffic can flow equally smoothly in both directions, the financial and other institutions must be capable of doing the right thing with the resources that come across the borders.

But what is the right thing, and what kinds of resources are desirable? The commentators stress the importance of attracting flows with long maturities, to lessen the vulnerability of financial systems, and in the right denomination (that is, preferably in local currency), to avoid undue stresses within the exchange rate regime. But one could also read between the lines a consensus on what should *not* be done. It is true that budgetary surplus is desirable, but fine tuning of fiscal policy might not be. Instead, a medium-term strategy is called for. The consensus on the right thing to do with the resources, once acquired, is simple and straightforward: invest and save.

The question then inevitably arises, how can countries promote investment and saving? The panel offers a broad menu of measures, all of them well considered: pension fund reforms, tax and regulatory incentives, and of course, as Domingo Cavallo emphasized, a stable economic environment. A stable environment with a policy emphasis on fighting inflation is the sine qua non for lengthening the planning horizons of savers and giving them incentives to save more.

If there is a point of disagreement, it is in regard to the sterilization of capital inflows. There are certain circumstances in which sterilization is not viable, and others in which it is not only viable but desirable. How does one tell them apart? One basis of discrimination is the time factor: changes deemed to be temporary are more likely to be sterilized than those that appear to be permanent. Temporary and permanent are, of course, in the eye of the beholder, and they are difficult to tell apart ex ante. Larry Summers' advice is therefore sound: be conservative, assume the worst. Policymakers who assume that outflows are permanent and that inflows are temporary are bound to err in the direction of safety.

Should developing countries impose capital controls? The question remains open. However, I agree with Larry Summers that one should not disband the fire brigades in the hope of discouraging smoking in bed. If one is worried about capital movements, the easy, tempting—and wrong—answer is to impose controls. Controls avoid all the costs of capital flow volatility, but they also avoid all the benefits of capital inflows. A good analogy is that the way to avoid automobile accidents is not to block all the roads, but rather to widen them—and perhaps also to require all drivers to fasten their seatbelts.

Jacob Frenkel is Governor of the Bank of Israel.

Part II
ACHIEVING STABILITY IN LATIN AMERICAN FINANCIAL MARKETS IN THE PRESENCE OF VOLATILE CAPITAL FLOWS

Liliana Rojas-Suárez and Steven R. Weisbrod

Introduction

The recent international debt and foreign exchange crisis in Mexico has demonstrated that volatile international capital flows can have severe consequences for domestic financial markets. Sharp increases in domestic interest rates, associated with the crisis, have weakened the solvency of short-term borrowers, raising concerns about the stability of the Mexican financial system. Recent sharp movements in capital flows have had similar, although less dramatic, effects on the financial systems of a number of other Latin American economies, particularly Argentina.

Since the late 1980s, many Latin American countries, aware of the need to strengthen their domestic financial systems, have undertaken a wide range of initiatives to liberalize their financial markets. These have included lowering reserve requirements on bank deposits, removing interest rate controls on bank assets and liabilities, and reducing asset allocation programs (government-mandated rules specifying the percentage of total credits that must be directed to specific economic sectors). The objective of these reforms has been to encourage reliance on market forces, rather than on direct bank controls, to allocate credit efficiently.

Policymakers in Latin America remain committed to liberalization, but are concerned about the effects of large and highly volatile capital flows into and out of the region on their financial systems. After all, the reforms have been instituted in financial systems in which the dominant instruments are short term: bank deposits that fund short-term bank loans and short-term government and central bank securities. Thus, when official interest rates are increased to defend an exchange rate parity or to fight inflationary pressures, the entire system is affected.

This raises the question of whether the Latin American economic environment is simply too unstable to rely on market forces and policies that exert only indirect control over financial markets. Advocates of reform believe the harmful effects of relieving banks of burdensome regulatory requirements can be greatly diminished by strengthening bank supervision, such as through capital requirements and improved procedures for recognizing nonperforming loans. These supervisory controls, based on standards now in place in the industrial market economies of the Organization for Economic Cooperation and Development (OECD), are designed to prevent excessive growth of credit to risky borrowers.

Skeptics counter that market forces and supervision, although appropriate in the OECD countries, are not strong enough constraints in the more volatile financial markets of Latin America. For example, while supporters of financial liberalization have advocated reducing reserve requirements and eliminating interest rate controls on banks to promote efficient allocation of bank credit, skeptics argue that high reserve requirements are necessary in Latin America to control the growth of bank credit. Thus, those who argue for nonmarket constraints on banks emphasize the importance of controlling liquidity growth, while those who support liberalization stress the need to control risk in the financial institutions that issue liquid liabilities.

This basic argument about the ability of liberalized financial markets to cope with the vagaries of Latin American economies extends to other areas as well. For example, is it prudent to encourage the expansion of nonbank capital market institutions as alternative suppliers of credit to governments and corporations? Can the central bank's balance sheet act as a stabilizing force in the financial market? Should the central bank accumulate large stores of international reserves relative to liquid assets, and if so, how should these reserves be financed? Should the central bank and other bank supervisors have the authority to provide credit to the banking system? Should the central bank permit the banking system to offer U.S. dollar deposits in competition with local-currency deposits?

The stresses generated by the recent Mexican financial crisis have been felt throughout Latin America, both in countries that have liberalized their financial markets and in those that have not. The crisis thus provides a unique opportunity to evaluate whether financial liberalization has provided sufficient protection against volatile capital flows. One can examine, for example, whether supervisory standards modeled after those in the OECD countries have provided sufficient stability in the Latin American markets where they have been applied. If they have not, is the solution to return to more direct controls, or should indirect controls, such as supervisory monitoring of bank loan quality, be maintained but modified for the more volatile Latin American markets?

In addition to evaluating financial market performance, policymakers must evaluate their procedures for handling severe banking problems that might arise as a result of a crisis. Policies that make it clear that bank stockholders will lose their investment if their bank fails can deter excessive risktaking; rigorous enforcement of such policies in time of crisis is imperative for policymakers to maintain credibility.

The following section, "Achieving Stability in Latin American Banking Systems," considers whether reserve requirements have been effective in controlling excessive liquidity growth. It discusses the adequacy of bank supervisory standards, such as risk-weighted assets standards for bank capital, in controlling the expansion of risky bank credit, which often accompanies excessive liquidity expansion. It also discusses the role that dollarization plays in protecting the domestic financial system in the event of an exchange rate crisis. The third section, "Do Capital Markets in Latin America Contribute to Stability?" analyzes whether the development of domestic capital markets has improved or hindered policymakers' ability to isolate those markets from volatile capital flows. The fourth section, "Dealing with Banking Crises: Lessons from the United States and Chile," reviews the experiences of the United States and Chile in resolving their recent banking crises, and derives lessons on how cleanups can be managed to reduce long-term risk in the financial system. The final section offers some concluding remarks.

Achieving Stability in Latin American Banking Systems

Since the late 1980s, in tandem with comprehensive macroeconomic stabilization programs, many Latin American countries have embarked on major structural reforms designed to improve the capacity of markets to set prices and allocate resources efficiently. The restructuring of financial markets has been a crucial component of this endeavor. Here we consider the recent experiences of Argentina, Brazil, Chile, Colombia, Mexico, Peru, and Venezuela in liberalizing their banking systems. Major efforts in some of these countries have included reducing reserve requirements on deposits and eliminating interest rate controls on deposits and loans. Reserve requirements on domestic currency deposits are much lower today in many Latin American countries than they were in the late 1980s, and in some cases, the early 1990s. Mexico has completely eliminated its reserve requirements on bank deposits, Peru and Venezuela have reduced theirs significantly, and Argentina and Colombia have reduced theirs more modestly. In sharp contrast, however, reserve requirements in Brazil have increased significantly in recent years. Today, at least some bank deposit and loan interest rates are determined by the market in all seven countries.[1]

As noted in the introduction, a number of analysts question whether one can rely on market forces alone to promote stability in financial environments as volatile as those in Latin America. Does supervision alone provide sufficient stability, or are stronger policy tools, such as high reserve requirements, necessary to control excessive volatility in liquidity resulting from international capital flows? Looking at the varied experiences of the seven countries in our sample, we address three issues in this section. First, we examine whether countries with relatively high reserve requirements on bank deposits have had more success in controlling liquidity growth than those with relatively low reserve requirements. Second, we consider whether supervisory standards have been effective in reducing the growth of high-risk credit that often accompanies rapid liquidity growth, and discuss how regulators can use a variety of market and accounting signals to assess the quality of bank portfolios. Third, we discuss how dollarization affects the banking system in a currency crisis such as the recent one in Mexico.

Deregulation and the Expansion of Liquidity: Are High Reserve Requirements a Good Idea?

A sudden reversal of capital flows out of a country can lead to doubts about the viability of its exchange rate regime and about the soundness of the domestic financial system. The loss of foreign exchange reserves that accompanies such an episode can be exacerbated if a large share of the economy's financial assets are held in liquid form such as bank deposits, which can easily be exchanged for international reserves. The rationale for high reserve requirements is that they facilitate the control of these liquid instruments. By reducing the interest rate paid on bank deposits and increasing the interest rate on loans, high reserve requirements lessen the attractiveness of both loans and deposits.

In addition, reserve requirements placed on banks are a funding source for the central bank—that is, they are a liability on the central

---

[1] For a summary of the process of financial reform in Latin America, see Westley (1993).

bank's balance sheet. High reserve requirements increase the size of the central bank's balance sheet relative to bank balance sheets; by reducing the stock of domestic money, high reserve requirements can increase the ratio of international reserves to bank deposits.[2] Supporters of high reserve requirements argue that a high ratio of international reserves to deposits provides protection against an attack on the currency, because if investors holding liquid assets in local currency should demand U.S. dollars from the central bank, their demands could be satisfied without disrupting local credit markets. Because the central bank has large dollar reserves, it does not have to resort to high domestic interest rates to defend the exchange rate.

Thus, while recognizing that reserve requirements are an inefficient tax on the banking system, proponents of high reserve requirements contend that they serve a prudential function by controlling liquidity growth. Opponents respond that the issue is not the relationship of bank deposits to international reserves, but rather the relationship of all liquid assets to international reserves; high reserve requirements, they maintain, encourage the substitution of other forms of liquidity for bank deposits.

We can test empirically whether high reserve requirements, by forcing a tight relationship between the quantity of liquid assets and the monetary base, are useful in controlling liquidity. If reserve requirements can control liquidity, countries with high reserve requirements will display a low ratio of liquidity to monetary base, and liquidity growth will be tightly constrained by growth in bank reserves.

Figure 1 plots the ratio of domestic-currency liquid assets to the monetary base[3] against marginal reserve requirements on demand (sight) deposits and on time deposits for the seven countries in the sample as of 1994. Liquidity for each country is defined broadly to include currency held by the public as well as both bank deposits and nonbank liquid assets held by the nonbank public (see appendix for the definitions of liquidity adopted by each of the seven countries). The major nonbank assets included in this definition are short-term government securities and liquid liabilities issued by the central bank.

---

[2] A special case is that in which the central bank operates as a currency board; then bank reserves can be created only if the central bank acquires foreign currency in an equivalent amount. If the central bank follows a floating exchange rate policy, it can freely determine the nominal quantity of bank reserves, but it can follow a policy of buying foreign currency at the market price for the bank reserves it creates.

[3] The monetary base excludes reserves held against foreign currency deposits at the central bank.

FIGURE 1. LIQUIDITY/MONETARY BASE RATIOS VERSUS RESERVE REQUIREMENTS ON DEMAND AND TIME DEPOSITS IN SEVEN LATIN AMERICAN COUNTRIES, 1994

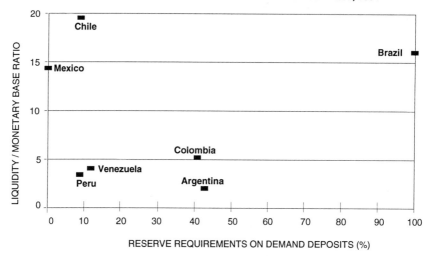

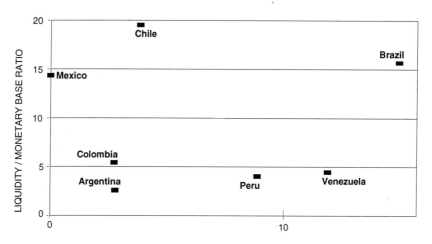

If high reserve requirements lead to a low ratio of liquidity to monetary base, the data in these two charts should roughly fit a downward-sloping curve. But as is evident from the figure, in 1994 there was no clear relationship between the liquidity-to-base money ratio and reserve requirements, for either demand or time deposits. For example, Chile's reserve requirements on domestic currency demand deposits are very similar to those in Peru and Venezuela, yet Chile has a substantially higher ratio. Colombia's reserve requirements on demand and time de-

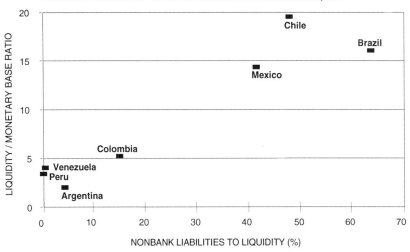

FIGURE 2. LIQUIDITY/MONETARY BASE RATIOS VERSUS RATIOS OF NONBANK LIABILITIES TO LIQUIDITY IN SEVEN LATIN AMERICAN COUNTRIES, 1994

posits are similar to those in Argentina, yet Colombia has almost three times the ratio of liquidity to monetary base as Argentina.

The reason for the lack of a definite relationship between reserve requirements and the ratio of liquidity to monetary base is evident in Figure 2, which depicts the relationship between the ratio of nonbank liabilities to domestic liquidity (on the horizontal axis) and the ratio of liquidity to monetary base. The ratio of nonbank liabilities to liquidity is highest in Brazil[4] and lowest in Peru. Clearly, the importance of nonbank money market securities is a more important determinant of the liquidity-to-base ratio than are reserve requirements. Government and central bank short-term securities are important components of liquidity in Brazil, Chile, and Mexico, which also have the highest ratios of liquidity to base, even though they have very different reserve requirement policies.

Anecdotal evidence suggests that even in the absence of open market paper, informal market arrangements evolve to undermine the effectiveness of high reserve requirements. For example, so-called *mesa de dinero* (wholesale money) markets, in which corporations buy and sell corporate receivables, have been large when high nominal interest rates and high reserve requirements make it expensive to hold demand de-

---

[4] The ratio for Brazil has been declining, however, since assets of the banking system have been expanding relative to the short-term government bond market. The data for Brazil in Figure 2 are for August 1994.

FIGURE 3. RESERVE REQUIREMENTS ON DEMAND AND TIME DEPOSITS VERSUS REAL LIQUIDITY GROWTH IN SEVEN LATIN AMERICAN COUNTRIES, 1993-94

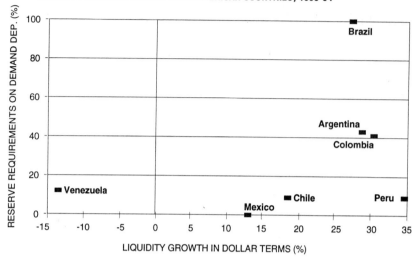

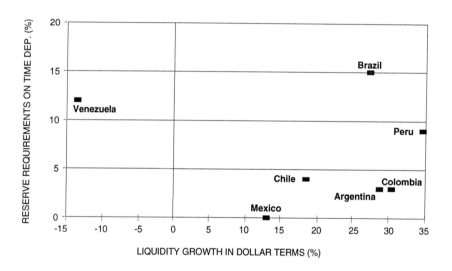

Note: Liquidity data are averages of real annualized growth rates for 1993 and the first six months of 1994 (data for Brazil are for the first five months of 1994). Real growth is defined as nominal growth adjusted for exchange rate changes.

posits. In Brazil, postdated checks are used in many transactions, reducing the need to hold demand deposits, which are subject to 100 percent reserve requirements. These informal markets rarely show up in the data on liquidity aggregates.[5]

Figure 3 demonstrates that on average, over 1993 and the first six months of 1994, there was no negative relationship between real annualized growth in domestic currency liquidity, defined as nominal growth adjusted for exchange rate changes, and reserve requirements on either demand deposits or time deposits.[6] Even though reserve requirements are a tool to control nominal rather than real liquidity growth, we measure that growth in dollar terms because of the disparity in inflation rates across countries, which distorts the data significantly. An interesting observation is that Mexico, with no reserve requirements, experienced decidedly moderate liquidity growth over the period.

The final issue to consider concerning the efficacy of high reserve requirements is whether they lead to a low ratio of liquidity to international reserves. There are several reasons why no such relationship need exist. First, as is evident in Figure 2, not all liquid assets are liabilities of banks. Second, central banks can use reserve requirements to fund domestic credit expansion, as happened in many Latin American countries in the late 1970s and early 1980s. As the experience of the debt crisis amply documents, this kind of credit expansion often leads to large losses in international reserves. Third, reserve requirements are only one liability on a central bank's balance sheet. Central banks can, and in several countries do, issue open market paper. They can also fund themselves with government deposits or long-term foreign liabilities.

The lack of a negative relationship between high reserve requirements and a low ratio of liquidity to international reserves is confirmed in Figure 4, using 1994 data. For both demand and time deposits, the level of reserve requirements has no discernable effect on the ratio of liquidity to international reserves. Mexico, with zero reserve requirements, had an exceptionally high ratio of liquidity to international reserves, but the evidence from the other countries does not suggest that a lack of reserve requirements was the cause of this situation.

---

[5] At major Brazilian banks, demand deposits represented about 13 percent of all deposits as of December 31, 1994, compared with ratios of 20 percent or more in most other Latin American countries. These data imply that Brazilians use substitutes for demand deposits for transaction purposes, as the anecdotal evidence suggests.

[6] Brazilian data are for the first five months of 1994, annualized. The negative liquidity growth in dollar terms observed for Venezuela is due to the devaluation of the bolívar in early 1994.

FIGURE 4. LIQUIDITY/INTERNATIONAL RESERVE RATIOS VERSUS RESERVE REQUIREMENTS ON DEMAND AND TIME DEPOSITS IN SEVEN LATIN AMERICAN COUNTRIES, 1993-94

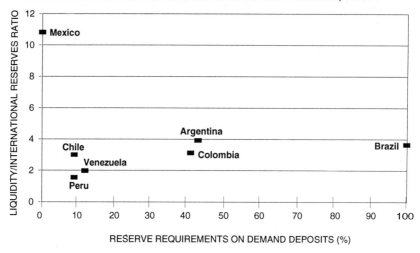

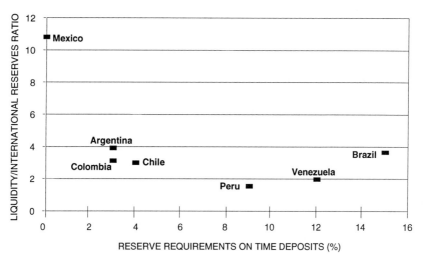

Bank Supervision as a Tool for Financial Stability

Supporters of financial liberalization have argued that improved techniques to supervise banks could replace direct controls on the banking system, such as high reserve requirements, for the purpose of maintaining control over reckless increases in bank credit. Unlike reserve requirements, which control only the amount of liquidity, bank supervision ideally affects both the quantity and the quality of credit expansion.

An important supervisory tool that has been implemented over the past decade is an international standard for the ratio of a bank's capital to its risk-weighted assets.[7] This standard, developed by the Bank for International Settlements (BIS), was instituted to ensure that banks internalize the cost of holding high-risk assets. The current internationally accepted minimum standard is a ratio of capital to risk-weighted assets of 8 percent, with half of that capital being equity capital.

In a number of industrial countries, investors' perceptions about the availability of public support for banks in a crisis prevents interest rates on bank liabilities, especially deposit liabilities, from fully reflecting the riskiness of banks' asset portfolios. In much of the industrial world, capital represents a more expensive source of funds than deposits, because the public commitment to bail out bank shareholders is not as strong as the commitment to bail out liability holders. Hence, by requiring banks to hold more capital against risky assets, supervisors raise the cost of funding such assets, which should lead banks to hold safer portfolios.

The imposition of capital standards on banks can also reduce the growth rate of bank balance sheets. Imposing capital standards on banks constrains asset growth. High capital ratios must be supported by high net income relative to assets. Thus, if banks are to grow without increasing leverage, they must neither bid aggressively for deposits nor bid for loan volume by reducing loan rates.

Risk-weighted capital standards have had a profound effect on banking behavior in industrial economies such as those of the United States and Japan. Regulatory authorities in the United States use these standards to determine whether a banking organization should be permitted to engage in new activities, and banks with deficient capital are not permitted to expand their assets. Investors in the United States have also adopted the risk-weighted measure in their own evaluation of the soundness of individual banks: the stocks of banks with low capital ratios trade at lower price-to-earnings ratios than those of other banks, partly because if a bank must issue new equity to meet the standard, the likely result, at least in the short run, is a dilution of earnings per share. Moreover, in Japan over the last few years, supervisory standards have

---

[7] This ratio is calculated by classifying assets by risk and assigning risky assets greater weight than less risky ones. For example, loans are typically given a 100 percent weight, whereas interbank deposits held with major banks in industrial countries are given a zero weight. For purposes of calculating the ratio, capital is defined as equity capital, including paid-in capital and the retained earnings account, which together are designated primary capital, plus certain preferred equity and subordinated debt issues, which are designated secondary or tier 2 capital.

FIGURE 5. BANK CAPITAL-TO-LOAN RATIOS VERSUS REAL LIQUIDITY GROWTH IN FIVE LATIN AMERICAN COUNTRIES, 1993-94

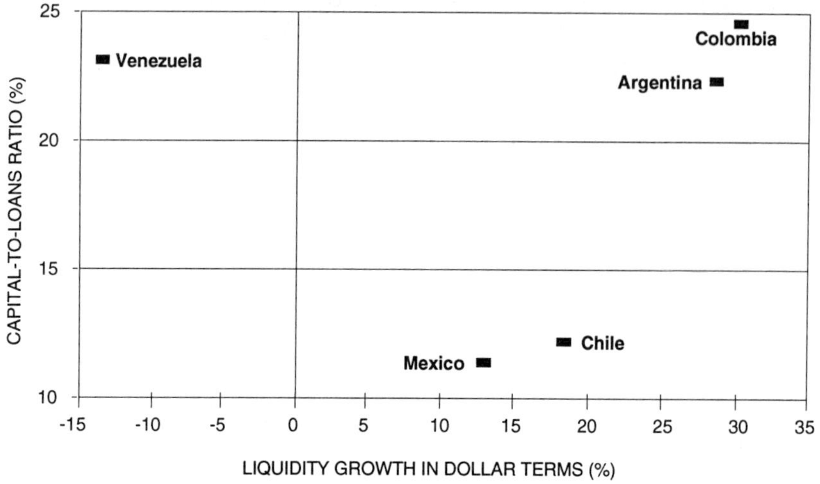

Note: Liquidity data are defined in the note to Figure 3.

led some major banks to restrict asset growth to comply with minimum capital ratios, which were severely affected by the asset deflation of the early 1990s.

At least five of the seven countries discussed in this section have adopted the capital standards of the BIS. A comparison of these countries' experiences allows us to ascertain whether these standards have proved effective in Latin America.

A rough indication of the effectiveness of capital standards across countries is whether banking systems with higher capital-to-loan ratios, used here as a proxy for the capital-to-risk-weighted-asset ratio,[8] have experienced slower real growth in loans, and therefore slower real growth in liquid liabilities of the financial system, than have countries with lower ratios. Like reserve requirements, bank capital ratios directly affect the growth of bank deposits; however, they also indirectly affect liquidity growth because banks are an important funding source for money market instruments in Latin America. Our interest is in measuring the effectiveness of both reserve requirements and capital ratios in controlling

---

[8] There are several justifications for this simplification. An important one is that the market for subordinated debt and preferred shares is small in most Latin American countries, and credit risk is likely to be the major risk facing the banks. For example, the capital-to-loans ratio in Mexico was 11.7 percent as of September 1994. This was fairly close to the 10 percent ratio calculated by the Bank of Mexico.

liquidity growth, because investors can sell any liquid instrument at its face value for international reserves.

Figure 5 depicts the relationship between capital-to-loan ratios and real liquidity growth; liquidity is again defined as growth in local currency liquidity adjusted for exchange rate changes, averaged over 1993 and the first half of 1994. Clearly, higher capital-to-loan ratios have not been associated with slower growth in liquidity in dollar terms.

Effective bank supervision, however, including evaluation of the adequacy of bank capital, requires a detailed analysis beyond these aggregate data; it is imperative to understand the condition of individual banks. It is difficult to determine from aggregate ratios whether capital standards are succeeding in restraining the growth of risky assets, or whether some banks are improperly reporting their capital-to-loan ratios. In accounting terms, bank capital is merely the difference between the value of a bank's assets and that of its liabilities. If assets are improperly valued, bank capital gives an inaccurate picture of the quality of bank balance sheets. A major determinant of the value of bank assets in Latin America is the percentage of bank loan portfolios comprising nonperforming loans. Hence, to evaluate the quality of a bank's capital-to-asset ratios, we need reliable evaluations of the quality of its assets.

One might begin by using additional accounting ratios provided by banks to the supervisory agencies—for example, the ratio of loan-loss reserves, also referred to as provisions, to nonperforming loans.[9] Among the ten largest U.S. commercial banks at the bottom of the recession in 1991, nonperforming loans amounted to 8.2 percent of all loans. The ratio of loan-loss reserves to nonperforming loans for these banks was about 45 percent. In Japan, the reported share of nonperforming loans at city banks (the eleven largest nonspecialized Japanese banks) was about 4 percent as of September 1994, and the ratio of loan-loss reserves to nonperforming loans was about 37 percent.

Among Latin American countries, these ratios vary widely. In Mexico as of September 1994, the ratio of nonperforming loans to loans was 10.5 percent, and loan-loss reserve coverage was 41 percent for the banking system as a whole. In Chile, for the banking system as a whole as of November 1994, 1.4 percent of outstanding loans were nonperforming, and loan-loss reserve coverage was 157 percent. For Colombia as of

---

[9] In the United States, the item on the income statement for provisioning against nonperforming loans is called "loan-loss provisions," whereas the corresponding balance sheet item is called "loan-loss reserves." Also in the United States, loan-loss reserves appear as a contra item on the asset side of the balance sheet. Other countries use different conventions.

September 1994, the ratios for large banks were 5.2 percent and 26 percent, respectively. Thus, whereas both of the Mexican ratios as of September 1994 were similar to the U.S. ratios in 1991, the Chilean and Colombian ratios of nonperforming loans were very low, even when compared with those of U.S. banks in more prosperous years. The loan-loss reserve coverage ratio in Chile was exceedingly high, whereas in Colombia it was below the level that U.S. banks attempt to maintain.

Again, however, aggregate ratios mask the behavior of individual banks. Some banks might hesitate to declare loans nonperforming because to do so implies a commitment to increase loan-loss reserves, which must be financed at the expense of net income. Since lower net income means lower retained earnings, which become part of the capital account, an increase in nonperforming loans indirectly reduces bank capital. Hence, if banks have incorrectly reported their nonperforming loans, they are probably overstating their capital-to-loan ratios.[10]

Thus, to obtain an accurate view of the quality of bank portfolios, one must often look beyond both reported capital-to-asset ratios and reported loan-loss reserve coverage ratios.[11] In addition, to assess the accuracy of these ratios, policymakers must consider the signals that the market is sending. For example, banks that are bidding more aggressively for funds than other banks might be taking on more asset risk to cover higher deposit costs. In 1985, the interest rate on deposits paid by U.S. money center banks (a group of very large banks headquartered in New York, Chicago, and San Francisco) was 70 basis points (0.7 percentage point) higher than the average for all U.S. banks. This spread increased over the next few years, so that by 1990 the money center banks were paying 168 basis points more than the average for deposits.[12] In 1990, nonperforming loans amounted to 6.85 percent of the loan portfolios of money center banks, compared with 4.19 percent for all banks.[13]

As discussed above, assessment of individual bank behavior is at the heart of bank supervision. To facilitate this analysis, supervisors of-

---

[10] For example, the Japanese definition of nonperforming loans (loans with past-due interest of sixty days and loans to bankrupt firms) is not as stringent as the U.S. definition (loans with past-due interest of thirty days).

[11] For a discussion of the difficulty of using capital-to-asset ratios as a measure of soundness of Latin American banks in the 1980s, see Rojas-Suárez and Weisbrod (1994).

[12] The decrease in spreads was not due to differences in the level of market interest rates, which were similar in 1985 and 1990. In addition, interest rates fell between 1990 and 1991, yet the spread remained substantially above the 1985 level.

[13] The financial problems of the money center banks in the early 1990s are illustrated by the fact that the largest, Citicorp, required a large infusion of capital from a private investor to maintain its ratio of capital to risk-weighted assets at the regulatory minimum.

ten use peer group analysis, classifying banks according to the type of customer they serve. For example, a commonly used segmentation is wholesale versus retail: banks whose clients are mostly large businesses versus those whose customers are mostly consumers and small businesses. Supervisors have certain expectations about how each peer group will behave. If the accounting ratios of individual banks within a group deviate significantly from those of their peers, this can be an early sign of trouble.

In Latin America, wholesale and retail banks might be distinguished by their ratios of demand (sight) deposits to assets, with wholesale banks having higher ratios.[14] Wholesale banks usually receive less income from interest relative to their assets than do retail banks, for two reasons: the default rate on retail loans is often higher than on wholesale loans, and noninterest expenses per loan are usually higher for retail loans (acquisition, servicing, and distribution costs per dollar of asset generated are higher in retail than in wholesale banking). In developing countries, there is often a third reason: a large portion of wholesale bank funding is through demand deposits, which have no or very low interest expenses.[15] Retail banks should have higher net interest margins (NIM)—the difference between interest income and interest expense—relative to assets because the risk is greater and noninterest expenses are higher.

Mexican banking data illustrate the value of peer group analysis in spotting problems in a banking system. Table 1 compares several averages of balance sheet and income statement ratios for selected wholesale and retail banks in Mexico, as well as for the banking system as a whole. The ratio of demand deposits to total deposits at wholesale banks was 37 percent, whereas at selected retail banks it was only 16 percent. Both interest income and interest expense as a percentage of assets were substantially lower at wholesale banks, as is typical in most Latin American markets. However, in contrast to these expected findings, the ratio of NIM to assets for the retail banks was below that for the wholesale banks. This suggests that the retail banks are not being adequately compensated for the risk in their loan portfolios. In addition, the selected retail

---

[14] In contrast, in the United States, where large corporations hold repurchase agreements (purchases of government securities with agreement to resell the next day) for their liquidity needs rather than holding demand deposits, wholesale banks often have a lower ratio of demand deposits to total deposits than do other banks.

[15] Data for several countries strongly suggest that demand deposits are a substantially cheaper source of funds for banks than other liabilities. In the United States, interest expenses at wholesale banks are higher than at retail banks because wholesale banks have a relatively low ratio of demand deposits to total liabilities, and their other deposits pay market interest rates.

TABLE 1. BANK BALANCE SHEET AND PERFORMANCE INDICATORS IN MEXICO, SEPTEMBER 1994
(PERCENT)

Indicator	All banks	Selected wholesale banks	Selected retail banks
Composition of assets (shares of total)			
Cash	2.1	1.9	1.5
Repurchase agreements	9.1	7.0	28.3
Loans	55.1	56.1	41.5
Securities	20.2	20.7	16.5
Composition of deposits			
Demand deposits as a share of total deposits	26.1	36.7	16.4
Performance indicators			
Loan-loss provisions as a share of nonperforming assets	40.7	41.9	55.2
Nonperforming loans as a share of total loans	10.5	10.7	11.3
Capital-to-loan ratio	11.4	13.2	9.1
Interest income/assets ratio	11.6	10.6	13.8
Interest expenditure/assets ratio	8.1	6.6	11.2
Net interest margin/assets ratio	3.5	4.0	2.7

Note: Income items are accumulated through September 1994 and are not annualized.
Source: Comisión Nacional Bancaria, *Boletín Estadístico de la Banca Múltiple*, September 1994.

banks' aggregate capital-to-loan ratio is lower than for the industry as a whole, implying that these banks have riskier portfolios than the average bank. These data indicate that wholesale banks in Mexico are less vulnerable than retail banks, a finding consistent with press reports on the types of Mexican banks reportedly facing financial difficulties since the onset of the recent crisis.

Analysis of individual banks is a powerful tool for effective bank supervision; yet supervisors, even when well informed about banking difficulties, face a major dilemma. On the one hand, they need to take prompt action against weak banks, but on the other hand, public announcements about problem banks might generate fears about the soundness of the entire banking system. These concerns can often be reduced by thorough and continuous disclosure of the condition of individual banks, to help the public distinguish between solid and weak institutions before matters build to a crisis.

Another problem is that, in spite of supervisors' best efforts to monitor the risks that banks are taking, supervision can be evaded, as can high reserve requirements. Banks can set up highly leveraged subsidiaries in offshore banking centers to hold risky assets. They can also increase their off-balance-sheet commitments, such as standby letters of credit,[16] in the domestic market. Hence, to be truly effective, supervision must be done on a consolidated basis and cover all bank commitments, whether on or off the balance sheet.

Financial Stability and Dollarization

A number of Latin American countries, as part of a set of policies governing the structure of their financial systems, permit banks to offer dollar-denominated deposit and loan accounts. In an exchange rate crisis, these policies can relieve pressure on domestic financial systems by allowing redenomination into dollars to avoid an attack on international reserves.

For such a safety valve to work, however, two conditions must obtain. First, reserve requirements on dollar deposits must not be greater than those on domestic currency deposits. Otherwise, an attempt to redenominate deposits and loans will result in a shortage of reserves and a rise in short-term interest rates.[17] Second, investors must be willing to hold their dollar-denominated deposits in the domestic banking system. If, for example, investors decide instead to place their dollar deposits in U.S. banks—that is, if there is a run on the local banking system—dollarization is likely to have the same outcome as a run on international reserves without dollarization. As investors leave the local banking system, they will demand international reserve assets to deliver as payment for deposits in New York. Among the seven countries in our

---

[16] Standby letters of credit commit a bank to pay off a credit obligation if a borrower loses access to nonbank sources of credit. Under BIS standards, standby letters of credit are included as an item in risk-weighted assets even though they are an off-balance-sheet commitment.

[17] If reserve requirements on dollar deposits are greater than those on domestic currency deposits, the shift to dollar deposits will require banks to deposit more funds with the central bank. The banks, moreover, must present dollar assets to the central bank to obtain the needed reserves. If the central bank will accept for discount only U.S. Treasury bills or dollar deposits in U.S. banks, then banks must reduce their loan-to-deposit ratios to obtain the required liquid assets; the result will be to raise the interest rate on both dollar and domestic currency loans. If the central bank is willing to discount dollar-denominated local assets, the interest rate on dollar deposits will rise, as investors perceive an increased risk that dollar deposits in local banks might not be converted into dollar deposits in New York banks. The latter point assumes that the shift toward dollar deposits results in a change in the mix of central bank assets held against banks' reserve accounts.

sample, five permit dollar deposits (Argentina, Chile, Colombia, Mexico, and Peru), and of these, only Argentina places the same reserve requirements on foreign currency deposits as on deposits in domestic currency. The other four place higher reserve requirements on foreign currency deposits.

The shift from local currency deposits to dollar deposits is most likely accompanied by a shift out of the local market when investors have doubts not only about the value of the local currency, but also about the soundness of the local banking system. In other words, if a country has a weak banking system, it is unlikely to avoid a currency crisis by adopting a policy of dollarization even if it maintains the same reserve requirements on dollar-denominated and domestic currency deposits.[18]

Although dollarization can be an effective weapon against a currency attack in the presence of a sound banking system, its benefits carry some costs. Most important of these is that the central bank loses seigniorage[19] profits. In cases of extreme dollarization, the central bank's ability to provide lender-of-last-resort protection to the country's banks is also diminished, since banks would need mostly foreign rather than domestic currency to deal with short-term liquidity problems.

Do Capital Markets in Latin America Contribute to Stability?

In the 1970s and early 1980s, much of the capital flowing into Latin America took the form of loans from foreign banks. In contrast, the two major sources of capital inflows in the 1990s were the return of flight capital by residents and the purchase of bonds and equities by foreigners, most notably through U.S. mutual funds. This leads to the question of whether foreign interest in capital market instruments issued by borrowers in developing countries has stimulated the growth and deepening of domestic capital markets, especially in long-term bonds and equities, which are considered stable sources of international capital. We conclude that in most countries, long-term capital markets remain thin. We also consider the extent to which the development of short-term securities markets, mostly for government and central bank liabilities, has affected the ability of policymakers to cope with variable international capital flows.

---

[18] Rojas-Suárez and Weisbrod (1994).
[19] These are profits that the central bank earns on assets it acquires when it puts money it has created into the market.

TABLE 2. NEW EUROBOND ISSUES AS A SHARE OF NEW DOMESTIC CREDIT IN SIX LATIN AMERICAN COUNTRIES, 1993–94
(PERCENT)

Country	Weighted-average share
Argentina	34.2
Chile	0.0
Colombia	16.7
Mexico	31.6
Peru	2.1
Venezuela	2.9

Note: Data are weighted averages for the first half of 1993 and the first half of 1994.
Sources: *International Financial Review, Global Financing Directory*, June 1993 and June 1994; International Monetary Fund, *International Financial Statistics*, February 1995; Banco Central de Reserva de Perú, *Nota Semanal*, February 1995.

The Development of Long-Term Capital Markets

Table 2 presents long-term Eurobond issuances as a share of total new domestic credit at financial institutions in six Latin American countries; the shares are calculated as weighted averages over the first half of 1993 and the first half of 1994.[20] The data indicate that domestic credit is a far more important source of funds than Eurobond issues. Even in Argentina and Mexico, where Eurobond issuance was relatively high, access to the market was limited to government institutions and a few large firms. It is noteworthy that Chile, whose financial system is more diversified than those of other Latin American countries, issued no Eurobonds in the two periods cited.

Table 3 takes a similar look at new equity issues in six countries in our sample: for each country, the table summarizes the ratio of new equity issues to new credits extended by the banking sector to private borrowers from the beginning of 1993 to the latest date for which data

---

[20] The weighted averages were constructed by summing new bond issues in the two periods for the numerator and summing new credit created in the two periods for the denominator. This is superior to a simple average because it increases the impact on the average of the period with the greatest credit expansion.

TABLE 3. RATIOS OF NEW EQUITY ISSUES TO GROWTH IN PRIVATE DOMESTIC CREDIT AT BANKING INSTITUTIONS IN SIX LATIN AMERICAN COUNTRIES, 1993–94
(PERCENT)

Country	Ratio
Argentina	46.2
Brazil	0.8
Chile	27.0
Colombia	0.2
Peru	4.8
Venezuela	0.6

Note: The last date in 1994 for which data are available varies by country.
Sources: Central bank bulletins (see References).

data are available. The data indicate that Argentina, where new equity issues equaled 46 percent of the increase in bank credit to the private sector, has been in relative terms the most active market for new issues in the recent past. It should be noted, however, that most of the new equity issues in Argentina during this period were the result of privatizations, representing one-time placements of equity. In addition, Argentina's equity market lacks breadth: three firms accounted for over 50 percent of stock market capitalization. New issues were also a significant share of financial institution credit in Chile, at 27 percent.[21]

Table 4 presents a breakdown, by type of instrument, of the volume of trading on securities exchanges for five of the seven countries in December 1994. Chile is the only market with active trading in a long-term bond, namely, a central bank security with indexed principal. In contrast, exchange trading in Colombia and Mexico is dominated by money market instruments. In Brazil, short-term securities are not included in the exchange data. Brazil does have trading in short-term securities markets—in fact, short-term government and municipal government securities are a large component of liquidity in that country—but that trading does not take place on the São Paulo stock exchange. Equity trading dominates securities exchange activity in Brazil and Peru. However, as Table 3 indicates, equity markets are a very limited source of finance in these two markets, since the volume of new issues is quite small relative to private credit supplied by banking institutions.

---

[21] The denominator in the Chilean figures excludes pension fund investments in private bonds.

TABLE 4. COMPOSITION OF INSTRUMENTS TRADED ON STOCK EXCHANGES IN FIVE COUNTRIES, DECEMBER 1994
(PERCENTAGES OF TOTAL)

Instrument	Brazil	Chile	Colombia	Mexico	Peru
Stocks	53.3	6.1	8.1	4.1	76.6
Money market instruments	0.0	40.6	74.2	64.2	23.4
Fixed-income instruments	0.0	53.2	17.1	31.7	0.0
Futures and options	46.2	0.0	0.0	0.0	0.0
Other	0.4	0.1	0.5	0.0	0.0

Note: Data for Brazil and Mexico are for November 1994.
Source: National stock exchange publications (see References).

- This evidence indicates that most domestic financial credit needs in Latin America are met through short-term bank loans and short-term government and central bank securities. The failure of significant long-term markets to develop can be explained by analyzing interest rate spreads among some dollar-denominated Latin American issues and U.S. Treasury notes and bonds of similar maturities. For example, these spreads in Mexico and Argentina are some 400 basis points higher than for an equivalent U.S. Treasury instrument. In Brazil, the spread has exceeded 700 basis points, as shown in Figure 6. These data appear to indicate that investors perceive Latin American instruments as riskier than corresponding U.S. instruments.

It is important to note, however, that the Chilean central bank issues long-term debt in the domestic markets that is actively traded on the Santiago stock exchange. The principal is indexed to inflation; the bonds yielded about 6.25 percent in real terms in early 1995.

It is sometimes argued that indexation of principal, whereby the nominal value of the borrower's debt increases with inflation, encourages the development of long-term capital markets. An alternative form of indexing, floating rate long-term debt, forces the borrower to pay increasing nominal interest if inflation rises, even as the real principal declines. Private lenders will prefer principal indexing only if they believe that borrowers' ability to repay is unaffected by the inflation rate; in a highly inflationary environment, lenders cannot be certain that borrowers' income streams are indexed to the same extent as their debt contracts.

Indexation might also serve a useful purpose in lengthening the maturity of government bonds when lack of full credibility about an-

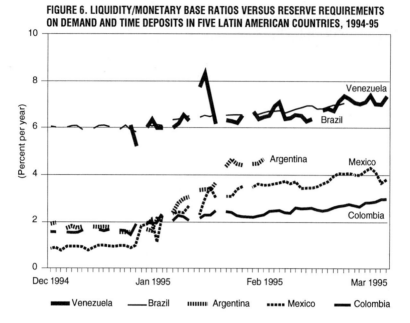

FIGURE 6. LIQUIDITY/MONETARY BASE RATIOS VERSUS RESERVE REQUIREMENTS ON DEMAND AND TIME DEPOSITS IN FIVE LATIN AMERICAN COUNTRIES, 1994-95

nounced policies leads to high ex ante interest rates. If, however, the government's commitment to low inflation is not upheld, the increases in principal payments that must occur under indexation will exert pressure on the fiscal accounts, raising inflationary pressures further. Hence, indexation in a highly inflationary environment cannot serve the purpose of increasing the maturity of public or private assets.

The Consequences of Short-Term Securities Markets

As indicated above, the evidence suggests that most of Latin American capital market formation has been concentrated at the short end of the maturity structure, especially among government and central bank securities. In industrial country markets, especially the United States and the London-based Euromarkets, the growth of short-term securities markets, including those in government and private paper, has provided competition to banks, reducing bank interest rate margins and increasing incentives for risktaking in the banking system.

In New York and London, bank interest rate margins have been reduced by money market competition, because depositors have viewed nonbank short-term paper as a close substitute for bank deposits. Hence, for example, in the United States, the spreads among rates for term Fed-

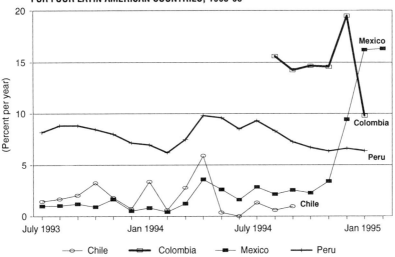

FIGURE 7. DIFFERENCE BETWEEN BANK DEPOSIT RATES AND INTERBANK RATES FOR FOUR LATIN AMERICAN COUNTRIES, 1993-95

eral Reserve (Fed) funds (the interbank rate), commercial paper, and bank certificates of deposit (CDs) are at most a few basis points. As indicated in Figure 7, which plots the difference between bank deposit rates and interbank rates for four countries in the sample, this arbitrage does not occur in Latin American countries, where interbank rates are significantly above deposit rates.

A common explanation for the lack of arbitrage in Latin American countries is that bank deposits are subject to reserve requirements, which depress deposit interest rates, whereas interbank loans are not. However, such arbitrage took place in the United States between term Fed funds and CD rates prior to 1991, at a time when there were reserve requirements on CDs. Among the countries represented in Figure 7, Mexico has no reserve requirements on bank deposits, yet the spread remains positive. Chile actually places higher reserve requirements on interbank funds than on deposits, and there, too, the spread is positive.

This lack of arbitrage has several consequences, some favorable and some unfavorable. Depositors receive low interest rates on their savings, and borrowers pay high interest rates. However, the wide spread, if well managed, could enable banks to accumulate capital and other reserve funds to protect themselves in a crisis. Even if banks' high net interest margins support high noninterest expense rather than capital buildup, in a crisis they can reduce noninterest expenses to help cover defaulting loans. For example, it has been reported that Banamex, the largest Mexican bank, has cut employment by one third during the current crisis.

In addition, as is evident from the spread between deposit and interbank rates in Mexico, during a crisis this spread increases. As long as banks can keep deposit rates below the interbank rate without losing too many deposits, they might also be able to keep loan rates low to keep borrowers solvent.

Thus, although the isolation of banking systems from international capital markets has its disadvantages in normal times, it can also help stabilize domestic financial markets during a crisis. In contrast to some industrial countries where arbitrage between banks and nonbank money market instruments prevails, the structure of financial systems in Latin America gives central banks some freedom to raise interest rates on open market paper to defend the currency without exposing bank borrowers to the full force of those interest rates.

Dealing with Banking Crises: Lessons from the United States and Chile

Despite their best efforts to put in place policies that will strengthen the domestic financial system, authorities must always be prepared to deal with financial crises. Crises of relatively large magnitude, requiring government assistance to private banking institutions, have occurred in both industrial and developing countries over the last ten to fifteen years. Even in the absence of a major banking crisis, regulators often must devise support systems to maintain the solvency of individual banks. However, if banking problems accelerate into a crisis, regulators need to be prepared with policies that protect the public, efficiently allocate financial losses, and strengthen the financial system.

This section reviews selected bank bailout experiences—in particular, those of the United States and Chile—to determine how programs might be designed to strengthen incentives for sound banking. The experiences of these two countries demonstrate that bailout programs are most successful under two major conditions. First, public credit to distressed institutions must be made conditional on a realistic appraisal of the loan portfolio: only borrowers that have a reasonable possibility of returning to solvency should be eligible for additional credit, and unpaid interest payments on loans to truly insolvent borrowers should not be accrued by issuing new loans. Second, equity holders and large liability holders must absorb the costs associated with the bailout to the full extent of their investment. These investors must realize that bank shareholders cannot enjoy the high returns of risktaking while benefitting from government protection against losses.

Extricating a banking system from a crisis, however, is a delicate balancing act. Selected liability holders in failed banks—usually small depositors—must be fully or partially paid off. The presumption is that small depositors are unable to evaluate the riskiness of financial institutions and hence cannot be held responsible for their failure. The public interest is also served by bailing out small depositors. The system as a whole is preserved by avoiding the danger that small depositors in solvent banks who lack information about their condition might withdraw deposits when they see that small depositors at other banks are losing money.

Nevertheless, the behavior of small depositors can make bank rescue efforts more costly. For example, failing institutions will often offer very high interest rates to attract insured depositors, then use these funds to roll over unpaid interest on nonperforming loans as well as to fund new high-risk loans.[22] The most efficient way to prevent insolvent institutions from continuing to pay high interest rates to attract insured deposits is to close these institutions and pay off the insured depositors as soon as the institutions become insolvent. Carrying out such a policy will usually require an upfront commitment of public funds, because the assets of insolvent institutions and of the deposit insurance system are usually not sufficient to pay off insured depositors. In practice, it has been the unwillingness of banking authorities to provide immediate public funding to close down such institutions that has led to bad policy.

In a number of countries, the resolution of severe banking crises has had two distinct phases. As the discussion below illustrates for the United States and Chile, the first phase, which often aggravates the crisis, involves attempting to provide bailout funds for institutions that have, in fact, failed. In the second phase, which is a reaction to the mistakes of the first, policymakers institute programs to close failed institutions to prevent them from taking further risks. This phase also includes efforts to strengthen the remaining institutions, such as enhancing supervision, improving loan workout programs, and more strictly enforcing capital requirements.

---

[22] When a bank is failing, stockholders and bank employees often have nothing more to lose by betting the bank on a high-risk venture, since their equity has already disappeared. A recent example of this is the Barings case, in which traders doubled their bets in the hope of extricating themselves from positions that were already in the red.

The First Phase: Bailout Attempts

In the United States, the banking crisis involved thrift institutions—depository institutions, such as savings and loans and savings banks, that have traditionally taken deposits from consumers to fund residential mortgage portfolios. In the late 1970s and early 1980s, the net worth of these institutions was impaired by an increase in interest rates on short-term deposits, which had been used to fund long-term home mortgages originated at substantially lower long-term rates. Regulators attempted to deal with the problem by providing aid to institutions with impaired capital, under the assumption that, with time, the institutions would be able to solve their problems.[23] This strategy was adopted because public officials did not want to provide the funds necessary to close down the institutions and pay off insured depositors.

Rather than buying time, however, the policy encouraged institutions with zero or negative net worth to take additional risk. Finally it became clear that attempting to keep these failed institutions in business was only adding to the eventual cost of resolving the crisis.

---

[23] Until the early 1980s, these institutions paid regulated interest rates on short-term deposits insured by a government-sponsored agency, the Federal Savings and Loan Insurance Corporation (FSLIC), and held long-term fixed-rate mortgages as assets. In the 1970s, as inflation rose to extremely high levels by U.S. historical standards, money market mutual funds developed, providing savings and loan depositors with an alternative savings instrument that paid market interest rates. As depositors withdrew their funds, regulators tried to protect the thrifts from this outflow with emergency liquidity assistance. But the outflow was too severe to manage, and as a result, the government began a period of gradual deposit deregulation.

Although deregulation stabilized the flow of funds into the savings and loan industry, it also placed tremendous pressure on the net interest margins of individual institutions. Many held long-term mortgages made at the low interest rates that had prevailed before the 1974 oil crisis, and they could not afford to fund them at the high interest rates of 1979 and 1980. As a result, the net worth of many institutions became impaired.

The first attempts to deal with the crisis were made through the Federal Home Loan Bank System, the system for providing liquidity to thrifts. Federal Home Loan Banks issued net worth certificates to institutions with impaired capital. These certificates, which appeared as assets on the balance sheets of the savings and loans, effectively permitted them to transfer the risk of underperforming real estate loans to the Federal Home Loan Bank System. Because the net worth certificates raised the value of savings and loan assets relative to their liabilities, the net worth of troubled institutions increased, which permitted them to continue to operate under the regulations then in force. In effect, regulators permitted these institutions to expand their activities in the hope that they could solve their problems over time.

This relaxation of the limits on their activities permitted savings and loans with almost no market net worth to bid aggressively for insured deposits and take large risks on the asset side of the balance sheet. The owners of these institutions had little to lose by taking large risks. The problem was compounded by the fact that the mortgage market was becoming securitized, which reduced the spread between mortgage rates and other long-term securities.

The Chilean banking crisis of the early 1980s was of a different nature. Chile's banks had borrowed heavily from foreign banks and invested a large portion of their assets in foreign currency loans to domestic borrowers. When these borrowers could not repay the loans, the banks were in danger of defaulting on their foreign currency liabilities.

The rescue effort in Chile got off to a somewhat unpromising start. The central bank made extensive credit available to the banking system and to defaulted borrowers, without establishing programs to restructure defaulted loans. Net lending by the central bank to financial institutions increased from 7 percent of GDP in 1982 to over 16 percent in 1983. In addition, the central bank purchased nonperforming loans outright from the banks.

In rescheduling nonperforming foreign currency loans, the central bank offered borrowers favorable exchange rates, which created severe losses for the central bank: the central bank exchanged the borrowers' pesos for dollars with which to repay their loans, but gave them more pesos for their dollars than they could have obtained in the market. The losses absorbed by the central bank were not sustainable, and it was forced to abandon this phase of the rescue in early 1984.

The Second Phase: Implementing Sound Restructuring Programs

The second phase of the savings and loan crisis in the United States was signaled by a shift in philosophy: rather than make it easy for thrifts with no market net worth to continue operating, the government would seize these institutions. To this end, the Resolution Trust Corporation (RTC) was established and given authority to issue bonds to cover the cost of restructuring the industry.[24] The RTC also received some direct funding from the government. The agency's job was to close savings and loan institutions that had no market net worth and sell off their assets. Equity holders in the failed savings and loans lost all of their capital, and depositors holding large amounts in a single account lost the uninsured portion (everything over $100,000). Assets in savings and loans declined from $1,350 billion at the end of 1988 to $832 billion at the end of 1992.

In addition, several steps were taken to strengthen regulation of the thrifts. A new insurance agency was established under the direction of the more experienced supervisors of the bank insurance fund, the Federal Deposit Insurance Corporation (FDIC). Capital requirements were

---

[24] A separate agency, the Reconstruction Finance Corporation, was established to issue bonds and provide additional funds to the RTC.

strengthened, and the regulators were given authority to restrict the growth of any institution that did not maintain adequate capital. Restrictions were also placed on the ability of savings and loans to make leveraged loans on nonresidential real estate.

The savings and loan program established under the RTC had several advantages over the bailout program. Most important, the RTC program had adequate funds to force risky savings and loans into bankruptcy, which established the principle that owners of savings and loans faced the financial consequences of failure. By providing an injection of government money, the program reduced the burden on sound savings and loans of having to bail out the more risky ones. (Insurance funds for financial institutions typically are financed by charging insurance fees to member institutions. If the insurance fund is to bear the entire burden of a bailout, these fees must increase.[25])

In Chile, the second phase began in 1984, when the central bank assumed both the nonperforming foreign currency assets and the corresponding foreign currency liabilities of the banks. Under the initial agreement, the banks were forced to buy back the bad loan portfolios they had sold to the central bank over a ten-year period at the original face value of the loan plus accumulated unpaid interest at a below-market rate, rather than at a value determined by how the loan was performing. The central bank had few domestic resources to fund the bailout program, so it was financed substantially with foreign funds.

An important feature of the original agreement was that the banks remained in charge of administering the loan portfolios they had sold to the central bank, which meant that they retained responsibility for collecting loan payments and encouraging borrowers to remain current on their payments. Thus, the central bank avoided becoming directly involved in managing loans to the private sector.

The loan portfolios of some banks had deteriorated so badly that they could not be rescued. The government took over these banks, writing down the value of their equity by marking the assets to market. The shareholders of these banks suffered large losses. The central bank then offered the distressed banks for sale to the private sector. It was able to find buyers, but only by contributing capital. The sales agreements were written so that the central bank had first claim on the earnings of the

---

[25] Even with an injection of government funds, the debt of the savings and loan insurance fund is substantial. Recently a controversy has arisen as to whether surviving savings and loans should bear the cost of this debt. These institutions argue that the higher fees will place them at a competitive disadvantage to ordinary banks.

bank for dividend purposes. This policy had two beneficial effects: it demonstrated to an identified group of shareholders that, if banks fail, their equity holders will lose money; it also put new management in place who stood to lose their invested capital if they took the same risks as their predecessors.

It must be acknowledged that the plan was not a complete success. Some banks were unable to maintain their scheduled repayments to the central bank. In 1989, the ten-year payback period was extended indefinitely. Under this plan, banks can pay dividends only to preferred private shareholders (that is, the new shareholders, the *capitalistas populares*, who bought shares when banks were recapitalized in the mid 1980s) until their debt obligations are fulfilled. By the end of 1993, the banking system's obligation to the central bank was estimated to have reached $4 billion, or 10 percent of GDP. Despite these problems, however, the Chilean banking system appears to have been strengthened by the experience. It has maintained a disciplined growth rate, without relying on high reserve requirements for domestic currency deposits.

The experiences of Chile and the United States have been repeated in numerous countries. For example, in the recent banking crisis in Venezuela, regulators began by providing credit to weak institutions rather than closing them down. Every case demonstrates that implementing first-phase policies only makes solving the problem with second-phase policies more expensive. The obvious solution is to avoid first-phase policies altogether. However, opposition to spending the funds required to implement the second phase might not disappear until it is apparent to all stakeholders that the first-phase policies are unworkable.

Concluding Remarks

A major challenge facing policymakers in Latin America is the choice of policy instruments to maintain the stability of their financial systems in the presence of highly volatile capital flows. The current debate over how to devise effective tools centers around the question of controlling liquidity aggregates as well as risks to the financial system. The dominant financial assets in Latin American systems are liquid short-term paper issued by banks, central banks, and governments, which investors can readily sell if concerns develop about the exchange rate and/or the financial system. Faced with a financial structure of this kind, some analysts have recommended adopting policies aimed at controlling liquidity growth through reserve requirements as a way to protect Latin

American financial systems from volatile capital flows. Opponents of regulation, however, have argued that the emphasis on control of liquidity growth neglects the importance of managing risk in maintaining a stable financial system.

The evidence presented in this paper is that policy tools applied at the aggregate level—whether reserve requirements on bank deposits or capital standards on banks—show no consistent relationship with the growth of liquidity aggregates across countries, largely because there are too many formal and informal avenues by which investors can evade the policies. Hence we suggest that the fundamental challenge to policymakers is to devise supervisory tools to control financial risk which can be enforced at the level of the individual institution. This set of tools includes imposing balance sheet and income reporting requirements on banks, comparing reported accounting ratios across banks, and checking accounting data against market signals of financial institution risk. By judging the behavior of individual banks against comparable institutions and market benchmarks, supervisory standards can be made credible. Strong financial institutions, monitored by professional supervisors, will increase their balance sheets prudently. As a result, liquidity aggregates will grow at a sustainable pace. Finally, in the wake of a crisis, resolution procedures must be designed to place the cost of excessive risk on those parties most responsible for taking that risk.

Liliana Rojas-Suárez is Principal Adviser, Office of the Chief Economist, Inter-American Development Bank. Steven R. Weisbrod is Consultant to the Chief Economist, Inter-American Development Bank.

References

Banco Central de Chile. 1994. *Boletín Mensual.* Santiago (December).

Banco Central de Reserva del Perú. *Nota Semanal.* Lima (various issues).

Banco Central de Venezuela. 1994. *Boletín Mensual.* Caracas (August).

Banco Central do Brasil. 1994. *Boletim.* Brasilia (October).

Banco de la República (Argentina). 1994. *Boletín Estadístico.* Buenos Aires (December).

Banco de la República (Colombia). *Revista del Banco de la República.* Bogotá (various issues).

Banco de México. 1995. *Indicadores Económicos.* Mexico City (February).

Bloomberg Information Network Data Base.

Bolsa de Bogotá. 1994. *Boletín Trimestral.* Bogotá (October).

Bolsa de Bogotá. *Boletín Mensual Estadístico.* Bogotá (various issues).

Bolsa de São Paulo. 1994. *BOVESPA.* São Paulo (December).

Bolsa de Valores de Lima. 1994. *Informe Mensual.* Lima (December).

Comisión Nacional Bancaria. 1994. *Boletín Estadístico de la Banca Múltiple.* Mexico City (September).

International Finance Corporation. 1994. Emerging Markets Data Base. Washington, D.C. (December).

International Financial Review. 1993. *Global Financing Directory* (June).

International Financial Review. 1994. *Global Financing Directory* (June).

Rojas-Suárez, Liliana, and Steven R. Weisbrod. 1994. "Financial Market Fragilities in Latin America: From Banking Crisis Resolution to Current Policy Challenges." IMF Working Paper WP/94/117. International Monetary Fund, Washington, D.C. (October).

Westley, Glenn D. 1995. "Financial Reform in Latin America: Where Have We Been and Where Are We Going?" Inter-American Development Bank, Washington, D.C. (January).

Appendix: Definitions of Monetary Aggregates

Country	Liquidity	Bank deposits
Argentina	Total monetary resources in domestic currency plus total deposits in foreign currency	All deposits in domestic and foreign currency less document acceptances
Brazil	M4 (as defined by the central bank)	Demand deposits plus savings deposits
Chile	M7 (as defined by the central bank)	M3 (as defined by the central bank) less currency in circulation
Colombia	Total credit-generating liabilities of the financial system plus nonmonetary liabilities of the central bank held by the public	Total credit-generating liabilities of the financial system less currency in circulation, bonds of financial corporations, *cédulas* BCH, and commercial financing firms
Mexico	M4 (as defined by the central bank)	M2 (as defined by the central bank) less currency in circulation and bank acceptances
Peru	Money plus quasi money (of the banking and nonbanking systems, in domestic and foreign currency)	Liquidity less currency
Venezuela	M3 (as defined by the central bank)	M2 (as defined by the central bank) less currency in circulation

Notes: M=a certain category of money supply (demand, time savings, etc.), as defined by each country's central bank.
Cédulas BCH are a type of security issued by Colombian mortgage banks.

COMMENTARY TO PART II

L. Enrique Garcia

The economic environment in which Latin America now finds itself has five important characteristics. First, there is an undeniable shortage of savings in the region. Second, the region's capital markets are thin. Third, most Latin American countries and institutions, except some of the largest, lack steady, assured access to international capital markets. Fourth, there is a particular lack of long-term funding for the region's development. And finally, most of Latin America is engaged in an ongoing process of economic reform, which provides the context within which all matters relating to capital flows should be discussed.

The paper under discussion contributes to a debate now underway between those who would rely on supervision to maintain equilibrium in the financial system, and those who question such reliance and instead would impose or reimpose direct controls on the system. The issue of managing capital flows has both macroeconomic and microeconomic dimensions and is part of a broader discussion of the viability of market economies in Latin America and of whether there are viable alternatives to market-oriented reform.

In my view, there is no viable alternative to the model of reform on which Latin America has embarked. Market-based reform is undoubtedly a difficult road to development, and obstacles are sure to be encountered. But those obstacles should not be an excuse to revert to the old regime of controls or to delay fundamental changes. No set of reforms, however well conceived, can simply be implemented once and for all and then left to run by itself; like even the best luxury automobile, reforms need periodic maintenance. Capital flow volatility, financial system difficulties, and bank failures are inevitable in an open market economy: participants must be as prepared to lose as they are eager to win.

Opponents of reform, however, will try to use these patches of turbulence as an excuse to derail reforms altogether, and even supporters will sometimes lose their nerve. Privatization, for example, is popular with everyone in principle: investors get a new vehicle for profit, consumers get lower prices and better service, and the government gets rid of a major drag on its finances. But sometimes privatization fades in popularity when participants discover they have to play by the rules of the market. Likewise in banking: banking systems fail not because of what does or does not happen in the capital markets, but because of a lack of sound institutions, a lack of supervision, and the fact that many banks improperly involve themselves in nonfinancial activities through related loans.

Macroeconomic conditions in the industrial countries—high interest rates, for example—will, to be sure, have an impact on capital inflows, but by themselves they do not explain volatility in those flows. More important is the policy framework adopted by policymakers in developing countries to stabilize financial flows across their borders. First of all, they should avoid running excessively large current account deficits. When capital flows to a deficit country dry up, that country has no choice but to undertake a painful adjustment. Second, developing countries should adopt and adhere to coherent macroeconomic and microeconomic policies. There is no substitute for proper fiscal management, for profound liberalization, for clear rules of the game for investors, and for competition in all markets. This requires a free flow of information, with transparent administrative procedures and appropriate legal frameworks. At the same time, one should encourage domestic saving and development of capital markets. There needs to be an adequate mix of capital flows of different types and maturities: more direct investment, more medium- and long-term bonds and lending. In the absence of the proper institutions and transparent stock markets, countries should exercise care when they encourage short-term foreign portfolio investment, and they should work more on developing national capital markets that will stretch maturities. Privatization and pension fund development are two instruments that can contribute to this process.

The paper examines whether it is worthwhile to maintain high levels of reserves to control liquidity and thus exert better control over capital flows. Whether such a strategy will succeed depends on the coherence of policy as a whole. For example, a country that keeps abundant reserves but operates a loose monetary policy will usually end up transferring the extra liquidity thus absorbed from the financial markets to expand the fiscal deficit. On the other hand, a country that uses a high level of reserves to back up its monetary base, or, as Argentina has done, to support the banking system, is more likely to see its effort succeed.

Some will argue that reliance on bank supervision is feasible only in developed economies. I would agree that Latin American financial markets are not yet mature enough, but it is necessary to start somewhere. Supervision that goes beyond monitoring of macrofinancial variables to a more detailed analysis of individual banks and their risks is a step in the right direction.

As the paper also reminds us, financial crises are not peculiar to developing countries. What we learned from the collapse of the U.S. savings and loan system, as well as from the crises in Chile, Venezuela, and elsewhere, is that it is futile to try to save banks that are effectively bankrupt. The correct approach is to determine fairly and honestly which banks are sound and which are not. Those that are not should be shut down and their small depositors paid off; those that have a chance to survive, as determined by clear performance criteria, can be given some financing and support.

The paper argues that dollarization can be an effective policy, as the cases of Argentina, Bolivia, and some other countries illustrate. But dollarization is not a panacea: the outcome depends on the economic fundamentals. In the wrong macroeconomic environment—in the presence of large fiscal deficits, for example—the introduction of dollar accounts in parallel with domestic currency accounts will only cause private savers to rush to the central bank to purchase dollars, precipitating a run on reserves.

At the beginning of the recent Mexican crisis, market participants did not differentiate among Latin American economies but instead cut off the flow of capital to all indiscriminately. But as David Mulford says in his comment, we are now seeing investors beginning to discriminate among those countries that have better fundamentals and those that are not trying as hard as they should to solve their problems. It is important to recognize, however, that the gap that has emerged in the supply of long-term resources is wide and that most countries in Latin America will not regain full access to such resources for a long time.

Institutions such as the World Bank and the Inter-American Development Bank should play a catalytic role in supporting private capital markets to enhance the flow of long-term resources to Latin America. But because private firms are restricted from direct access to the financing provided by these institutions (with the exception of the International Finance Corporation), the privatization of so many former state enterprises cuts off the main external source of their long-term funding and obliges them to replace those resources in the capital markets. Under these circumstances, new programs are needed to provide insurance against political risk for long-term loans from private financial institutions. However, very few countries are eligible to issue medium- and long-term instruments in those markets. Under these circumstances, multilateral institutions should expand their financing to the private sector, either directly or through enhancements, in order to be consistent with the role that the private sector is expected to play in the market-oriented development process. The Andean Development Corporation, which I am honored to head, is playing precisely that role, since it can act directly with the private sector. In fact, about 60 percent of the ADC's two billion dollars in loan and equity approvals in 1994 went to the private sector without sovereign guarantees.

In conclusion, I believe that countries in Latin America should observe what I call the five Cs. The first is consistency: a country's policies, macroeconomic and microeconomic, must be in harmony with each other. The second is continuity: leaders and governing parties will inevitably change, but countries should strive to maintain a national consensus on the basic policies and philosophy of government that will survive these changes. From the first two Cs proceeds the third, credibility, and these three in turn are the anchor that attracts and holds domestic and external capital—the fourth C—within the economy. The first four Cs lay the foundation for the fifth, namely, growth (*crecimiento* in Spanish), which is the precondition for improved social well-being and sustainable development.

L. Enrique García is President of the Andean Development Corporation.

David Mulford

As a former practitioner in financial markets who later served in government, and who is now a former government official once again participating in world financial markets, I bring, perhaps, a unique perspective to the relationship between governments and markets. In my comment on the paper, with which I mostly agree, I shall therefore address the broader picture of international financial markets and the relationship between them and the economies of Latin America, and, in so doing, try to put the issues raised in that paper into perspective.

The 1980s are often regarded as a lost decade for Latin America. Certainly it was a decade in which Latin America lost the access to global capital that it had enjoyed in the period immediately preceding: after 1982, capital flows to the region fell to extremely low levels, and what remained came mostly from the official multilateral institutions or was squeezed out of those commercial banks that found themselves trapped in the system by their own past lending. I remember speaking frequently during those years on the question of how those countries might once again make themselves attractive competitors in the world capital markets sweepstakes.

From the vantage point of the 1990s, however, the 1980s might not have been a lost decade after all. It was then that the foundation was laid for the reform and restructuring of the Latin American economies whose success is now so evident. We all know that reform takes time, that it is a political process, and that it must begin at home, with the people of the reforming country shouldering most of the burden themselves. The 1980s were the years when the political decisions were made to restructure the Latin American economies in the direction of markets.

Discussions about capital flows are often couched in terms of whether flows at a given moment are too small, as they were in the 1980s, or too large, as they might have been more recently. But the deeper question is, Why did capital flows to Latin America resume so dramatically in the early 1990s after a long period of starvation? The answer, of course, is the region's renewed attractiveness in terms of investment opportunities, and it is a great tribute to the people and the political systems of Latin America that they were able to inspire so rapid and complete a turnaround. But markets—both in finance and in goods and services—also had a lot to do with Latin America's recovery. Markets are now a reality in the region, yet they are still insufficiently respected and understood by policymakers. Whatever the cause of the violent fluctuations in capital flows that the region has experienced, the fact remains that the capital inflows provided by world financial markets are good for a country, if it has the policies in place to attract capital and if it knows what to do with that capital when it arrives.

Four things are important to remember as the debt crisis of the 1980s recedes into history. The first is that the so-called Brady Plan, which was central to resolving the debt crisis, was in fact a market-based solution. By creating the financial instruments that are still in use today, it in effect securitized Latin America's commercial bank debt and provided a new vehicle for sustaining investment in the region. The second is that the resolution of the crisis was made possible and supported by a sustained policy reform and restructuring effort within the region itself. One cannot overstate how impressive that process was, in terms of both the strength of the commitment to change and the extent to which the economies of the region have been transformed. Latin America is quite simply a different place today than at any time in the past thirty years. The third point to remember is that intraregional trade and investment have also been transformed and have become a most important part of the overall picture. And the fourth point is that, with its recovery from the debt crisis, Latin America as a whole and certain countries in particular have become integrated into an ever-expanding, ever more sophisticated and fast-moving world capital market. The proof of this last assertion, ironically and unfortunately, is Mexico: until the recent crisis there, it was perhaps not widely understood just how deeply integrated certain developing countries have become into global capital markets.

The progress of capital market integration raises the associated issue of exchange rate regimes. In years past, when countries and currencies were largely isolated from the world financial system, the choice of an exchange rate regime was a relatively minor matter. Here, too, the Mexican crisis demonstrates that this is no longer the case.

In short, the debt crisis of the 1980s was resolved, and resolved in such a way that the composition of capital flows to Latin America changed radically. Meanwhile, the success of policy reforms in the various countries, together with a renewed flow of capital, was key to the further liberalization and growth of domestic capital markets.

In the context of these positive developments, however, the recent problems in Mexico demand that we step back and carefully reassess what has been accomplished so far. Clearly, although the Mexican economy was managed by experienced and professional people, serious policy mistakes were made. But were the mistakes so serious in and of themselves as to justify the ensuing violent reaction on the part of the financial markets? It is here that a market perspective is helpful. The Mexican officials apparently disregarded, or failed to understand, the likely reaction of the markets to their decisions, or they might not have fully appreciated the degree of integration into the world market that their country had achieved. That unfortunate miscalculation has had calamitous consequences not only for Mexico but for all of Latin America.

It is fair to say, in defense of the Mexican authorities, that markets sometimes overreact. But markets are also hungry for information, and are constantly reappraising each country's prospects. Often when they appear irrational they are in fact looking beyond the immediate concerns of the policymakers. Mexico's *tesobonos* (government bonds denominated in U.S. dollars), for example, were designed in such a way that they actually made Mexico more vulnerable when the crisis came. Their denomination in U.S. dollars, calculated to inspire confidence that they would maintain their value, in fact had the opposite effect when the markets realized that the peso could not maintain its level against the dollar, and therefore that Mexico might not be able to pay its dollar-denominated debt.

Whoever is judged at fault for the Mexican crisis, all of Latin America now finds itself in a situation of extreme volatility, and indeed, until the weeks just prior to this conference, the international financial markets were failing to discriminate among the situations in the different countries as well as they should have. Now, however, they are beginning to take a more careful look and to separate their assessments of different countries. There is reason for concern, however, that the wrong lessons will be learned from the recent turmoil. The sentiment is growing within some policy circles that dependence on foreign capital is too dangerous for a country to tolerate, and this could lead to the reimposition of controls. Conversely, however, with the announcement of the U.S.-Mexican support package, there is the danger that investors from industrial countries will come to expect a bailout when their investment in risky ventures in developing countries go sour.

To return to the question of why the market seems to have punished Latin America so severely for Mexico's mistakes, part of the answer is that the reversal of flows was already underway before the Mexican crisis hit at the end of 1994. Interest rates in the United States had risen very sharply during the year, and the recovery in Europe was competing for investors' attention as well. It is important also to distinguish between short-term and long-term flows. Most of the recent turbulence has been in the market for short-term capital; the long-term capital market—consisting in large part of major corporations making direct investments for strategic purposes—has remained relatively calm. This is a good sign, because it indicates that these important strategic investors continue to regard the fundamentals as positive. It is also a positive sign that lending by official institutions and commercial banks has diminished in relative importance.

Another explanation for the market's flightiness is the fact that new investors in Mexico found, when the market began to turn against them, that they lacked the liquidity they were accustomed to in markets elsewhere. This, of course, has nothing to do with the underlying performance of Mexican firms or the Mexican economy, but is simply a matter of traders panicking when they could not obtain stock prices at a moment's notice. There might also have been the fear as the crisis unfolded that history was repeating itself, that it was 1982 all over again, even though the situation was completely different. Finally, the dramatic intervention of the U.S. and Mexican authorities and the International Monetary Fund, however appropriate and timely, might have contributed to a sense of crisis and intensified the market's fears.

Events such as those in Mexico inevitably give rise to talk of providing some kind of safety net. But it should be obvious to anyone involved in financial matters—it is certainly obvious to the markets—that no safety net can be made big enough or strong enough to protect the entire global financial system from collapse should it plunge itself into a sudden decline. Perhaps more resources could be provided and some new operations set up within the international financial institutions to address certain specific concerns, but it is surely beyond the capability of even those institutions to insure all the vast flows of money and capital now circulating in world markets.

One lesson to be drawn from the recent events was noted by Domingo Cavallo: once countries decide to become integrated into world capital markets, they no longer have the same policy options. Devaluation in a country with open capital markets is not a through ticket to paradise. It hurts the very people whose capital one is seeking to attract. In such circumstances, a policymaker who devalues is very much like a politician who, once elected, reverses course and introduces legislation counter to the interests of his principal constituents. Devaluation remains a policy option, but one that countries that want to maintain capital inflows have to weigh much more carefully than before.

What lies ahead for Mexico, and what needs to be done? The first order of business, of course, is to stabilize the economy, and I believe that Mexico will do so, although it will take time. But the aftermath of the crisis should be a time of reflection as well as action. Investors need to reflect on the risks they are taking, which, in the case of Mexico, proved greater than they originally perceived. They also need to distinguish more clearly, in their investment decisions, among countries and among countries' policies. For their part, policymakers need to reflect on the nature of the financial markets on whose support they increasingly rely; to understand that the market makes its judgments from day to day—and judges severely; and therefore to be constantly aware of what is happening in those markets. Meanwhile, officials need to maintain the momentum of their reforms, and give particular attention to areas that have been somewhat overlooked. I have in mind especially those areas that relate to national saving, such as efforts to reform pension regimes. The authorities should also seek to identify policies that can afford some protection against interruptions in short-term capital flows. In addition, there is obviously going to have to be more discipline in the management of exchange rate regimes than was thought necessary in the recent past.

Finally, recent events underscore the need for policymakers to redouble their efforts to communicate with investors on a regular basis. Once a country has decided that it wants capital flows, that it welcomes them and will use them to build its economy, it must build transparency into all its policies and operations. Policymakers must be prepared, day in and day out, to communicate clearly and fully with the thousands of individual and institutional investors who make the decision every day whether to invest or to stay invested in the country's economy. Emerging markets are, after all, still only emerging: they offer great promise, but they remain unfamiliar territory for most investors worldwide.

In closing, although I said before that we should not look to official institutions to provide a safety net, that is not to say that I am opposed to improvements in surveillance—on the contrary, there is much to be done in that regard, and the international financial institutions ought to have more resources at their disposal to perform that function. Those resources must be real resources, not some elaborate paper shell game, and they must come from the major industrial countries, so that there will be no doubt whatsoever about the credibility of the institution put in charge of them, whether it be the IMF or some other organization. But, again, there are not enough resources in the world, and certainly not enough resources at the disposal of governments, to finance a safety net to protect every investor against every contingency. In the end, the stability of financial markets relies on policy discipline and on a keen awareness and sound understanding of how they work—not just on the part of policymakers and policy advisers in Latin America and elsewhere, but also on the part of the international institutions themselves, which have not always been as oriented to markets as they might have been.

David Mulford is Chairman of CS First Boston, Ltd. He was Undersecretary for International Affairs of the U.S. Treasury during the Bush administration.

Arturo C. Porzecanski

As the one panelist who has not had much of a career in the public sector, I will try to provide an unashamedly private-sector, free-market view on the questions raised in the paper. The first of these is whether policymakers can achieve stability in domestic financial markets by relying on supervision of financial institutions. My short answer to that is no. I think that inflows and outflows of capital have to be managed through macroeconomic rather than microeconomic policy.

The financial system is often treated like the messenger who is shot for delivering the message. Yet the financial system is the echo chamber of the economy; one regulates the sound by going to the source of the sound, not by tinkering with the echo chamber. Many of the pernicious effects of the volatility in capital inflows and outflows in Latin America would be much smaller if countries would take seriously two fundamental problems. The first is a shortage of domestic savings: without a larger pool of savings, Latin America is never going to have self-sustaining economic growth, or indeed sustained economic growth at all. For the second half of the 1990s, the countries of the region need to make improving the domestic savings rate a top policy priority.

Certainly, reducing public dissaving by cutting public sector deficits is an important first step, and one that most Latin American countries have taken. But Latin America has yet to take the second and quantitatively much more important step of increasing the pool of private savings. The successful Asian countries differ markedly from the only occasionally successful Latin American countries on this score. Latin America can point to one success story, namely, Chile, where the pool of financial savings was dramatically increased by privatizing the social security system. Similar reforms of pension systems elsewhere are, to be sure, not the only element in a strategy of raising domestic savings, but they are an important element and should be given high priority.

The second problem relates to the exchange rate regime: an important step in promoting the stability of capital movements is to move to freely floating exchange rates. There has been endless discussion in Latin America of the relative merits of different exchange rate regimes, and countries have tried all kinds. But seldom or never in recent memory has a free float, of the kind that the United States, Japan, and Germany now have, been attempted for any sustained period within the region. By a free float I do not mean one in which the central bank is indifferent to the level of the exchange rate, but one in which desired changes are effected through monetary policy, not by imposing or raising reserve requirements on banks or trying to maintain exchange rate movements within a band. The fixed exchange rate regimes now in place have encouraged capital inflows by encouraging all kinds of arbitrage operations. But, as we have seen, in other circumstances they have also encouraged capital outflows.

I believe that high reserve requirements only drive a wedge between deposit and loan rates, encouraging disintermediation, and transfer financial resources to the central bank, where one might question whether they are put to good use. The question of whether reserve requirements are effective or ineffective in protecting foreign exchange reserves against speculative attack is irrelevant under a floating rate regime: the best situation for a government is not to have to protect its reserves or the exchange rate from speculative attack. When emergencies arise, governments have to be able to focus on improving the fundamentals through fiscal and monetary policy while letting exchange rates move freely.

The paper also considers whether bank supervisors can effectively gauge the quality of bank balance sheets from the data currently available to them. I think that the quality of bank assets is notoriously hard to assess from reported data, and not just in Latin America. Needless to say, the same is true for nonbank securities companies. The best solution is for regulators to create risk-adjusted capital adequacy standards, enforce regulations that restrict lending to related entities, reduce large concentrations of risky assets, improve loan classification and provisioning procedures, try to reduce moral hazard dilemmas, and enhance the regulators' own supervisory capacity. While it is true that the data that bank supervisors have available to them leave much to be desired, that is no excuse for not doing what can and needs to be done.

With regard to dollarization, I believe that governments should neither encourage nor impede it. Savers and borrowers, and economic agents in general, should be free to specify contracts as they see fit. The paper also raises the issue of supervision of banking institutions that have subsidiaries overseas. That, too, is a familiar problem, the only solution to which involves international cooperation and common standards among supervisory authorities. Those are the keys to effective supervision, but they are not infallible ones, as we have recently seen.

Would the present crisis be less severe if the market for long-term funds, which remain scarce throughout the region, had been better developed? Again, I think that the only thing that would have diminished or prevented the present crisis is a higher level of domestic savings and a regime of freely fluctuating exchange rates.

My answer on the merits of indexation is the same as for dollarization: I think governments should neither promote nor impede it. Again, efficiency is best served by allowing financial and economic agents to enter into contracts according to whatever terms they see fit. Should bank supervisors encourage entry into banking? Yes, but they should also encourage exit from banking, to maximize competition and efficiency.

Finally, in discussing bank supervision in time of crisis, the paper makes a distinction between a first phase, in which policymakers take patchwork measures rather than face the need to close insolvent institutions, and a second phase, when the need to intervene becomes too obvious to ignore. On the grounds that foresight is never perfect, I believe it is impossible to expect that policymakers will skip the first phase and go straight to the second.

I think many governments have done a great disfavor to their countries by saying that banks will not fail. Bad banks should fail. The two really lasting solutions are, first, prevention, in the form of effective supervision, because prevention is always better than cure; and second, arrangements that allow failing institutions to achieve an orderly exit from the financial industry.

Arturo C. Porzecanski is Managing Director and Chief Economist at ING Capital Holdings.

CONCLUSION TO PART II

Jacob Frenkel

The commentators have addressed the issues raised in the paper from a market perspective, offering similar insights but with different emphases. David Mulford began by noting that volatile capital markets are a fact of life. We need to learn to live with them and to exercise the appropriate degree of selectivity.

The discussion also highlights the very real danger that policy solutions to economic problems sometimes end up killing the messenger—that is, they suppress the activity of the economic indicator that alerted policymakers to the problem in the first place. The market is a very delicate messenger, and it has a long memory. Policymakers should therefore handle it with the greatest care. This is especially true of all solutions that are basically inward looking, whether they be capital or other controls, daylight solutions that raise concerns about moral hazard, or some form of safety net. Every safety net has its holes. The policymaker's job is to determine where the holes are, and when things fall through, who will bear the financial burden. There really is no universal, all-encompassing safety net. The real question is, what policies need to be put in place to minimize the chance that safety nets will have to be used?

Is the present situation a repeat of the 1982 crisis? Clearly not. There are many fundamental differences between that episode and the present one. The number of market participants, for example, is much greater. In 1982 it was possible to round up a few bankers, appoint a steering committee, and work out a solution. Today capital markets are much more diverse, and much more dispersed, with hundreds of thousands if not millions of individual players.

The topic of devaluation, which was treated more fully in the context of the first paper in this volume, reappears in this discussion. The popular view is that devaluation provides an easy escape from external macroeconomic pressures. But David Mulford correctly pointed out that devaluation is not a magical solution, because markets pass judgment continually. As Albert Camus said, "We should not wait for the day of judgment since it takes place everyday."

The rationale for government intervention in financial markets is ultimately based on externalities. Safety nets are public goods; they not only protect those institutions unfortunate or incompetent enough to fall into them, but, by maintaining the financial system's integrity when those institutions fall, they benefit the stronger institutions as well. The proper form and extent of intervention are still matters for debate, and the comments represent both sides of the discussion. We were reminded that the raison d'etre of the international financial institutions is to create such externalities, and that at the very least, they should exercise an important surveillance function to ultimately reduce the need to use safety nets.

Surveillance, however, depends on adequate information. Transparency is important, as are regular updating of financial statistics and early warning signals. Once these are in place, the financial markets, which already know how to use this information, will be able to price financial assets accurately.

In the present financial environment, the nature of economic shocks is very different from what it was before. Whereas in the past, as Leonardo Leiderman has often noted, shocks tended to come by way of the terms of trade—that is, through the current account—today they come via the capital account, through interest rates, and they work their effects quickly and with a vengeance. How do countries deal with this exposure? The old approach was based on closedness, on import substitution. Today the preferred answer is openness and promotion of exports, together with an emphasis on domestic saving—in short, the East Asian model. The contest between these two opposing models remains undecided, but the Asian model seems to gain a little more ground each day. This relates to the question of credibility, which Enrique Garcia emphasized, because credibility is a matter not just of macroeconomic policy but of the overall strategy—inward or outward—of economic reform.

In opening their capital markets, countries should take care to move consistently in the same direction—not to take one step backward for every two steps forward. This is essential to maintain the confidence of foreign and domestic investors. To avoid proceeding by fits and starts, it is critical to plan each step carefully.

Exchange rate policy is relevant in this context. The problem is not that freely floating rates have not been tried—they have been. The problem is that the exchange rate system has not been invented that will save a country from policy mistakes. The question is, therefore, what exchange rate regime is most conducive to discipline in policymaking? The answer depends on the nature of the country's institutions, on its policymakers, on its history, and on the state of the inflation cycle. The danger is that in deciding on an exchange rate regime, policymakers will always think the grass on the other side is greener and will endlessly move back and forth between flexibility and fixity and thus fail to confront the real issue, which is how to establish discipline in policymaking.

Jacob Frenkel is Governor of the Bank of Israel.